AF541198

A-Z ECOLOGY

A-Z ECOLOGY

Dr. Ramesh Umrani

CENTRUM PRESS
NEW DELHI-110002 (INDIA)

CENTRUM PRESS
H.O.: 4360/4, Ansari Road, Daryaganj,
New Delhi-110 002 (India)
Ph.: 23278000, 23261597
B.O.: No. 1015, Ist Main Road, BSK IIIrd Stage
IIIrd Phase, IIIrd Block,
Bangalore - 560 085 (India)
Tel.: 080-41723429
Visit us at: www.centrumpress.com

A-Z Ecology

First Edition, 2009
ISBN 978-93-80106-14-4

PRINTED IN INDIA

Printed at Salasar Imaging Systems, Delhi-110035 (India)

Contents

Preface

Ecology is the scientific study of the distribution and abundance of life and the interactions between organisms and their natural environment. The environment of an organism includes physical properties, which can be described as the sum of local abiotic factors such as insolation (sunlight), climate, and geology, and biotic ecosystem, which includes other organisms that share its habitat.

The word "ecology" is often used more loosely in such terms as social ecology and deep ecology and in common parlance as a synonym for the natural environment or environmentalism. Likewise "ecologic" or "ecological" is often taken in the sense of environmentally friendly. The term ecology was coined by the German biologist Ernst Haeckel in 1866, when he defined it as "the comprehensive science of the relationship of the organism to the environment.

Ecology is usually considered as a branch of biology, the general science that studies living organisms. Ecology is a multidisciplinary science. Because of its focus on the higher levels of the organization of life on earth and on the interrelations between organisms and their environment, ecology draws on many other branches of science, especially geology and geography, meteorology, pedology, genetics, chemistry, and physics.

Author

Chapter 1

Scientific Place of Ecology

A wonder of science is the way in which all things interconnect and merge. The inconceivably miniscule and the incomprehensibly immense are of a continuum of complexity beyond the grasp of the ordinary mind. And yet, as Albert Einstein observed, perhaps the most profound wonder of all is the very comprehensibility of the natural world.

There is actually an astounding simplicity underlying the whole of the natural world. Nature is efficient, economic, and without excesses.

Occam's Razor Actually Functions: The simplest alternative explanation of a phenomenon is probably the most accurate. Undue complications (as opposed to the organization of complexes and complexity) are always more suspect in science than simplicity. The same rules work throughout the sciences.

The several natural sciences reflect an organization of nature that is quite apparent upon serious observation. This organization is composed of stepwise levels, each level representing a science that incorporates the principles and concepts ("rules") from those sciences below and establishing rules constraining those above, as described by the Theory of Integrative Levels, to be described below.

Despite these interconnections and merges across the lines established by scientists, there are those who assert that the study of any of the biological sciences, including ecology, cannot be undertaken without a firm foundation in basic physics and chemistry and all the disciplines of those sciences somehow relevant to biology.

Moreover, the abilities of a statistician, if not of a theoretical mathematician, are essential. While there is some truth to these assertions, they are not entirely acceptable, as will become apparent throughout this volume. Still, just as it is useful for the environmental professional to have a sensitivity to ecological principles, if not formal training in that science, it is incumbent upon the student of ecology to foster a sensitivity to the sciences contributing to ecology.

A valuable starting point for the nonscientist gingerly approaching ecology is some comprehension of natural science itself, of the scientific method, and of scientists and their culture. The three are really inextricably associated, especially as viewed from outside the scientific community.

SCIENCE AND SCIENTIFIC METHOD

Science is a systematic approach to comprehension of the natural world. Science is a unique human endeavor quite distinguishable in its methodology from all other approaches to understanding the cosmos and its components. Science can be defined as a body of facts and truths regarding the operation of general laws.

But scientific facts or truths are not so crucial as how they come into currency. The facts, hypotheses, theories, and laws of the natural sciences are continuously evolving by the cumulative application of the scientific method. Presumably, scientific explanations come ever closer to how the phenomena of the natural world are actually working and influencing one another. Throughout its history, science has grown through the gradual acceptance that events are not arbitrary but reflect an underlying order.

Sequences in scientific advance begin with sets of appearances that can be organized into laws. Somewhere along the array of laws a knot will occur. The knot is a point at which several laws intersect. Symbolically, the knot itself provides unity among the array. The methodology proceeds from observations of natural systems into the collection of data by empirical and experimental means. Empirical refers to observation of a system without intrusion into its workings, whereas experimentation involves some controlled alteration of the system of interest in order to clarify how it works. In either case, hypotheses, ideally universal in application, are tested. In principle, any number of analytical approaches may advance the hypothesis without ever fully confirming it, but it only takes one refutation to falsify or at least limit the universality of a hypothesis. Moreover, each hypothesis and its ultimate fate tend to be embedded in broad theory.

They do not rest in isolation from other parts of the natural world. Science advances through a combination of deductive and inductive reasoning with the constant application of probability and other statistical manipulations. Models of phenomena follow from the statistical confirmation and refutation of hypotheses. Over time, science is cumulative, despite momentary reversals and changes of direction. A striking picture of how science proceeds is derived from our changing view of the physical universe. There are three realities in this universe:

Matter, space, and time. To the classical (Newtonian) physicist, they were absolutes. At the turn of the century, special relativity demanded abandonment of absolutes in relation to space and time and required an acceptance of the equivalence of matter and energy. Quantum mechanics entered the picture only a few years later to demand yet another reexamination of these few realities of the universe and their interrelationships. All are expansions of one another by which scientific comprehension is advanced in clarity, outlook, or both.

Each scientist seeks the order to be extracted from appearances by looking for likenesses. Neither likenesses nor order will be immediately apparent but must be extracted. Thus, theory goes beyond any mere collection of facts. Historic discoveries result from insight into novel order. Sir Isaac Newton, for example, went beyond the pull of gravity on the apple. He realized that the force reaches beyond the tops of the trees and holds the moon in its orbit.

Michael Faraday closed the link between electricity and magnetism, and today we speak of the radiation of electromagnetic energy. James Clerk Maxwell linked electricity and magnetism to light. Einstein linked time to space and mass with energy.

LEVELS OF ORGANIZATION

Science is possible because we do live in an ordered universe in compliance with simple mathematical laws. The scientist studies, catalogs, and relates the orderliness in nature. The order is discernible wherever we look, from the galaxies spread across the universe to the internal workings of the atom. Still, it is useful to distinguish among the sorts of order. Some of it is found in the simplicity of regularities such as those associated with the clockwork of the solar system or the oscillations of a pendulum.

Another order is found in complexity (now called chaos and subject to mathematical analysis,) notably in the swirl of Jupiter's atmospheric gases or of incoming waves or in the organization of living beings.

Reductionism and holism are yet another kind of order. "Reductionism seeks to uncover simple elements within complex structures, while holism directs attention to the complexity as a whole." In reflecting this particular form of order, the Theory of Integrative Levels becomes a valuable guide into the sciences.

The Theory of Integrative Levels

Both scientists and students of their work who would put it to some particular use are greatly assisted by the very interconnectedness that is so important to our relationship to our environments, up to and including the biosphere. As John Muir observed, whenever we try to isolate a piece of the environment, we soon discover that it is "hitched" to everything else.

There are two ways to elucidate this point and demonstrate its significance to the environmental professional. One is what Eugene Odum calls the "Theory of Integrative Levels," presented in Figure. The other is a simplified version of systems analysis that can be illustrated as a series of black boxes, as shown in Figure. The Theory of Integrative Levels is no more or less than recognition that each level of organization within the natural sciences builds on those below in conceptual steps from pure mathematics to the planetary complex of ecosystems identified as the biosphere. In theory at least, mathematical relationships (quantification) are the foundation of all the natural sciences.

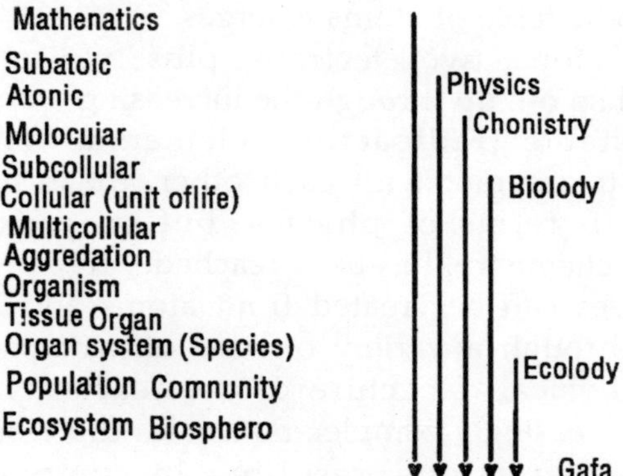

Fig. Theory of Integrative Levels: Levels of Organization among the Natural Sciences

Each level consists of features at least superficially amenable to qualitative description.

This narrative form can be rendered suitable to individuals outside the corresponding scientific discipline but is most often expressed within the discipline in terms of mathematics, most notably, but hardly exclusively, as statistical functions.

Quantitative relationships build in complexity within each level to a point where a new conceptual level comes into being. The new level, while dependent on these quantitative features, can no longer be fully understood through them alone. A qualitative alteration has occurred, with a corresponding emergence of a new level of organization. Significantly, no prior level is fully abandoned as complexity builds through the steps.

Hence, theoretical subatomic physics is very close to pure mathematics, but when all the subatomic interactions are in place, new features composing a new whole are discernible. Similarly, the functioning of certain of these phenomena becomes what is identified as protons, neutrons, and electrons. A single proton in specific interactive relationships with a single neutron becomes the next level: The combined phenomena constitute a hydrogen atom, the simplest of the chemical elements. Building on that simple foundation, the whole periodic table of atoms emerges.

Two protons, two electrons, plus two neutrons are helium, and so on, up through the increasing atomic weights to the unstable (radioactive) elements. When atoms commence to interact with each other, the results can be explained in terms of physics, but the next level of complexity, chemistry, has been reached.

Molecules can be created from atoms and from other molecules through a variety of reactions mediated by the internal (physical) structure of the reactants and their products. Increasingly complex molecules can result through molecular interactions, resulting in compounds and, ultimately, polymers composed of relatively simple molecular entities repeating in specific patterns.

Chemistry has traditionally been divided into inorganic and organic. Inorganic chemistry includes the properties of the elements and reactions of elements and simple molecules or compounds such as the metals, the halogens, gases, acids and bases, and water.

Organic chemistry is the study of reactions of the element carbon as it forms complex molecules in reactions with itself and especially hydrogen, oxygen, and to a lesser extent, nitrogen. While the next level of organization, sub cellular biology, requires the molecules and reactions of organic chemistry, it is ordinarily identified in terms of biochemistry and physiology. Both of these sciences, one claimed by chemistry, the other by biologists, imply a more elaborate complex than the organic chemistry upon which they are based. In essence, biochemistry describes the reactions and metabolic pathways giving rise to the physiology upon which the life of a cell or a multicultural organism depends.

Table. Selected Elements

Element	*Symbol*	*Commentary*
Hydrogen	H	Consisting of one proton and one electron, the simplest element; occurs as H_2 (the diatomic hydrogen gas, highly explosive), H_2O (water), a major component of organic molecules -- $(CH_2O)_n$
Helium	He	Only slightly more complex than hydrogen, consists of two protons, two electrons, and two neutrons; a gas (essentially non-reactive)
Boron	B	A trace element (essential to life) and heavy metal (toxic); a plant nutrient
Carbon	C	Occurs in three elemental forms -- diamond, coal and graphite; also occurs as gases -- CO (toxic carbon monoxide), CO_2 (essential to life), CH_4 (toxic methane); the basis of organic chemistry, in turn the basis of biochemistry, in turn the the basis of living material in the form of biochemical and physiological reactions
Nitrogen	N	Occurs as N_2 (the diatomic nitrogen gas), NO_2- (nitrite), NO_3= (nitrate), NH_3 (ammonia), NH_2 (amine); a major element of life
Oxygen	O	Occurs as O_2 and O_3 (the gases diatomic oxygen and ozone, respectively), H_2O, H_2NO_3 (nitric acid); an essential element in organic chemistry
Sodium	Na	Typical compounds NaOH (sodium hydroxide), NaCl (table salt)
Magnesiu m	Mg	A trace element and heavy metal; an essential component of chlorophyll that mediates the initial photosynthetic reaction

Physiology, then, implies a level of organization beyond, but based on, biochemistry, which in turn arises from organic chemistry. Because the complex of interconnected reactions known as life ultimately depends on the physics of atomic structure, it is unwise to neglect physics and mathematics in seeking to understand the processes of life. Nevertheless, it is both possible and essential to study and analyze biochemistry and physiology as independent levels of organization.

Similarly, it is both possible and essential to isolate for study and analysis each of the several levels of organization composing biological systems. The first of these levels is the cell itself, often identified as the unit of life. The highest level is the biosphere, although there are those who would argue that there is yet another biological system—the planet itself or Gaia. The intervening biological systems are the organism, population, community, and ecosystem.

From another perspective, there are levels of organization within the organism as well. Cells are organized into tissues, which are constituents of organs, which act together in organ systems. Each level lends itself to productive and informative study, but from the perspective of the organism, it is the complex of interactions that is important.

THE ENVIRONMENTAL SCIENCES

it is important to note that an arrow can be drawn from the point of input into the abiotic portion of the ecosystem or into any one of the populations representing the community of organisms. Thereafter, a veritable maze of arrows, some pointing in opposite directions, may proliferate before the one giving rise to the observable output is initiated.

It is also important to acknowledge that each population is a black box filled with individual organisms interacting with each other and with their environment. According to the Theory of Integrative Levels, the serious observer may well be required to work down through the layers of black boxes, all the way back to chemistry or physics for full comprehension of how something entered the ecosystem and was expressed as the measured output. At the same time, the serious observer will remain aware that both the input and output associated with the observed ecosystem are connected with the larger environment and, hence, potentially with the whole of the biosphere.

While it would be most difficult to isolate any of the sciences considered above as irrelevant to environmental professions, ecology is, nevertheless, the principal natural science serving environmental protection and natural resources management. Ecology commences where the individual organism ends.

The ecological sciences begin with the study of populations, communities, and whole ecosystems. But these biological systems obviously do not exist in a vacuum, scientific or otherwise. Hence, the physics and chemistry of geology are important, as are disciplines associated with the phenomena of the earth's atmosphere, hydrosphere, and lithosp here.

Too often ignored is the fourth dimension, time. No environmental science is complete without some reference to applicable units of time. Ecological phenomena occur over time spans as small as milliseconds and as large as geological intervals measured in millions and billions of years.

Also to be included in sound environmental planning are the social sciences, notably including economics, sociology, political science, and even psychology, human and otherwise. While it can be reasonably argued that ecology is the prerequisite to sound environmental science, none of these other considerations can be dismissed. There are times when one of them may legitimately supersede ecological considerations.

Life and, therefore, ecology are fundamentally chemical in nature. Chemistry depends on atomic structure, with special reference to the number and placement of electrons surrounding each atom's nucleus of protons and neutrons. The electrons, whose actual paths around the nucleus are not completely determinable, do occur in levels, determined by the number of electrons.

Chemical reactions occur between elements on the basis of the degree of completeness of each electron level or with the minute electric charge expressed by the atom. Covalent bonds, ubiquitous in the biological systems, represent the sharing of a pair of electrons between two atoms. Ionic bonds, also ubiquitous to biology, are related to attractions between positive and negative electrical charges. Some reactions are best understood as a combination of the two kinds of bonds.

There are a few atoms and in all likelihood no molecules or compounds that are not somehow significant to the processes of biological systems, whether the effect is negative or positive (or both). Table is a synopsis of the chemicals of particular significance to life.

Among molecules, liquid water may be designated the "universal mediator" of reactions, but carbon is better described as "the elemental jackof-all-trades." In properties, carbon lies halfway between the metals and the nonmetals. It can form compounds with almost all other elements-the property that causes carbon to be the unit of organic chemistry and hence the chemical basis of life.

Carbon is uniquely able to form covalent bonds with itself and with hydrogen, oxygen, and a variety of other elements and radicals. Carbon can form straight or branched chains of linked carbon atoms, cyclic compounds of five or six carbon atoms, and the so-called aromatic compounds based on benzene. Benzene is a six-carbon cyclic molecule, where every other carbon atom is bound to the next by two instead of one electron bond.

The simplest organic molecule is methane, a single carbon atom bonded with four hydrogens. The potable alcohol, ethanol, is composed of two carbon atoms, one with three hydrogens bonded to it and the other with two plus the hydroxy radical-OH.

It is possible to proceed stepwise from methane to essentially every known organic chemical. To do so in the laboratory is neither easy nor inexpensive, but it can be done whenever the reactions can be directed and controlled. The chemical wonder of living cells is that they have the means of synthesizing precisely the right chemical in the needed amount and at the proper rate to serve not merely their own needs but those of the whole organism.

Although it cannot actually be reduced to no more than the law of physics, life is not a matter of some novel force or other unprecedented natural laws. Life is actually a particular physicochemical complex that exhibits certain characteristics unique to it, as would be predicted by the Theory of Integrative Levels. All species of organisms have the innate, genetically mediated capacities to metabolize chemical substances, to excrete the wastes of metabolism and catabolism, to replicate themselves in a characteristic pattern identified as reproduction, to respond to environmental stimuli, to move, and to adapt.

There is a kind of "missing link" between living and nonliving in viruses, just barely crossing the line into life. Viruses are particles composed of nucleic acid and protein. They are acellular and lack metabolic processes.

They neither synthesize protein nor generate energy, but without either, viruses cannot grow. Yet they do reproduce themselves through altering the genetic code of the cells on which they are obligate parasites. Moreover, they can adapt and evolve. Many can actually recover from crystallization and, once reactivated, reproduce themselves.

Characteristics of Life:

- *Organization:* Specifically unique physical and chemical
- *Metabolism:* Utilization of raw materials from the environment to produce unique pattern of structural and physiological elements and energy distribution
- *Excretion:* Elimination of by-products of metabolism (wastes)
- *Reproduction:* Ability to replicate (make copies of self)
- *Sensitivity:* Responsiveness to environmental stimuli
- *Movement:* Ability to move; motility

Although the individual organism within a species may lack one or more of these capacities or lack the opportunity to perform one or more of them in a manner considered normal for the species, life is defined at the species level and more particularly distinguished from nonlife by all of these characteristics. All contribute something to ecological processes and must be taken into account for a full understanding of the phenomena of ecosystems and other ecological entities.

Qualitative vs. Quantitative Analysis Chemists and any biologist concerned with the effects on any biological system of chemical substances will expend substantial mental energy on identification and quantification of unknown substances, often in mixtures.

Take, for example, a vessel containing a clear fluid. In determining the nature of the fluid, qualitative analysis reveals the presence of gold, deuterium, hydrogen, and oxygen. The observer may next want to know something more practical: Is this mixture valuable? Is it toxic?

Is it hazardous in some other sense? To make any of these determinations, it is necessary to quantify the various components. In this case, gold may be present at a concentration of one part per million (1 ppm) and deuterium at 20 ppm. The mixture is neither valuable—at least as a practicable source of the precious metal—nor dangerously toxic—so long as it is not inhaled by terrestrial animals. It is mostly oxygen (850,000 ppm) and hydrogen (100,000 ppm). In fact, the fluid is seawater, the original source of life and present-day sustainer of life.

SCIENTISTS AND THE PRACTICE OF SCIENCE

From data collected through observation or experimentation or both, a hypothesis is formulated. In deductive reasoning, that hypothesis can be generally stated in terms of "If one thing is the case, then another is also" (If A, then B; A causes B.). Conclusions arising from the testing of narrowly focused hypotheses are expanded to increasingly general application through the process of inductive reasoning.

Needless to say, one scientist may not be engaged in every step of the scientific method. Moreover, since scientists are also human, insight and imagination play important roles in their individual professional lives, as do prejudices and biases. Nevertheless, science, at least over the long term, does tend to be self-correcting.

One way or another, the phenomena identified with the natural sciences will make themselves known, whether or not the individual scientist understands or believes in them. Although not in some simplistic straight lines resembling the national highway system, scientific trends in comprehension and accurate descriptions of the natural world build through the systematic aggregation of those portions each discipline advances. The aggregation is composed of the scientific method and constitutes science.

Ideally, each natural phenomenon is studied from a variety of perspectives, with a constant reaffirmation of the underlying hypothesis from different means of approach depending on different sets of assumptions and premises. Constant reaffirmation elevates one or more facts to constitute a theory that may eventually become a natural law or be dropped altogether.

For example, evolution in the form of neo-Darwinism is one theory of how species arise, stabilize, adapt to perturbation, and ultimately either undergo genetic change giving rise to phenotypes constituting a new species or become extinct.

This theory is subject to ongoing study, modification, and reevaluation as more and more information is incorporated. Moreover, the theory is further affected by other concepts of how species may be eliminated or how new taxons may arise.

Gravity, on the other hand, is an inescapable natural law. While the mechanism of gravity is not fully understood, the fact of gravity is universally accepted among scientists. It is crucial in research ranging from the physics of fundamental particles and associated forces to the phenomena of electromagnetic radiation to cosmology. Gravity also illustrates how a science evolves.

For Newton, gravity was a force, the one holding the planets in their orbits about the sun. In fact, Newton's mathematical formula for gravity can still calculate those orbits. For Einstein, gravity was curvature in spacetime. According to Einstein's mathematics, planets follow the shortest path in curved spacetime. In weak gravitational fields, the curvature in spacetime is almost perfectly flat, and Einstein's formulas give the same answers as Newton's.

Gravity's lesson expands upon an early one regarding scientific advance. A theory is good to the extent that it accurately describes a large class of observations on the basis of a corresponding model with few arbitrary elements and to the extent that it makes accurate predictions regarding future observations. Theories are always provisional, never proved. With each confirmation comes greater confidence, but each disagreement calls for modification, if not abandonment.

If the data generated by an individual scientist do not confirm or reaffirm the hypothesis being tested, the scientist has not failed. Instead, the hypothesis must be rejected—itself an advancement of relevant knowledge-or modified to accommodate the result. Science, like judicial rule-making, / proceeds in small increments, with only the occasional profound breakthrough and new direction, each increment flowing from one or more precedents.

Often, the mission behind the practice of science goes beyond comprehension of the natural world for its own sake. The basic purpose behind these formal processes is the ability to predict events on the basis of currently acceptable hypotheses as they coalesce into theory and take on the aspect of natural law. A system about which predictions can be accurately made can be subjected to control, a crucial element of the overall process for environmental professionals, whatever their setting in the private and public sectors.

In this context, it is essential for environmental professionals to accept that science is not a democratic process. Despite the fact that "respectable" science is accepted by the majority within a scientific discipline, the majority may in fact be wrong. It requires an astute observer to distinguish the inspired individual, another Einstein, from the "crackpot." Sometimes, substantial time must pass before the distinction becomes clear, with the majority proven correct or with the formation of a new majority. The scientific phenomena remain sublimely undisturbed by the whole process. They continue according to natural law in any event. They do not depend on human comprehension for their reality.

There are certain characteristics of scientific fact that must be taken into account by scientists and nonscientists alike. First and foremost, it is crucial to accept that the natural processes known collectively as science are real and themselves fixed in the natural world. They cannot be altered by scientists, technologists, or participants in the politicolegal system. They can only be increasingly understood, utilized, and applied. The better we comprehend the basics of the phenomena, the more accurately we can predict the result of an alteration—perturbation or intervention—in a system of specific interest. Such predictions then become the foundation of controlling either the result or the perturbation.

With that given, it is next crucial to appreciate the reality that all scientific "fact" is tentative. Future observations and experiments based on their predecessors will almost certainly cause resultant "facts" to be clarified, modified, or even refuted. Moreover, predictability in the scientific context is limited. It, too, expands but only with continued observation and experimentation, both of which require time that scientists may have but that environmental planners do not. In part because of the limitations of human knowledge and in part because of the tentative nature of scientific knowledge, predictability in specific circumstances is also limited, particularly at the interface of science with policy and law, a special kind of applied science. Moreover, the scientific reality itself is not subject to human control.

Finally, the phenomena science continually seeks to reveal transcend the work of the individual scientist, the discipline associated with each, and even the scientific endeavor itself. In short, no one can expect ever to have a full and complete understanding of the phenomena of a single discipline, let alone of the cosmos as a whole. Each of us works in a limited area subject to our own intellectual limitations superimposed on those of the practitioners of the field.

IMPLICATIONS FOR ENVIRONMENTAL PROFESSIONALS

Implicit to the processes of science is that every serious scientist is willing to abandon a theory when evidence contradicts it. Scientists, however, are known for their professional conservatism. Resistance to novelty is no flaw but ensures that novel notions be compellingly demonstrated if not proven outright. "Maverick notions" are ordinarily unreliable, albeit momentarily tempting.

The record of science chronicles only the innovations that have been shown to be right. Sooner or later, the wrong notions drop out of the collective known as science. Were it not for the conservatism, testing, and weeding-out process, science at large would be highly unreliable.

These tendencies and processes together ensure that the sciences reflect the natural world with increasing accuracy. Nevertheless, they are a major source of frustration for anyone who would apply science to practical problems demanding immediate action.

Also worthy of serious reflection by environmental professionals is that even mainstream scientific results may be counter to intuition and ordinary common sense. Everyday experience can prove to be an unreliable guide into natural phenomena, which are more accurately described through resort to the abstract or to the arcane realm of mathematical manipulation. Even more difficult than working with these idiosyncracies of scientific practice is conveying these points to the nonscientist policymaker or private citizen.

It can be reassuring to know that the processes of science demand the occasional "reality therapy" in maintaining a true course. The interpretations are adjusted and the accuracy of the maps into the natural world improved. For all their inadequacies, the maps science provides can be "perfectly reliable in important respects."

The trick, of course, is to know where reliability phases into misguided or simply wrong. Human nature is such that beliefs are relinquished only with the greatest reluctance.

Sometimes, it is nearly impossible to conceive experiments that could prove it wrong.

And yet, because of the process itself, science is self-correcting. Neither honest mistake nor fraud will go undetected indefinitely. All scientific research is open to critical examination and testing by others within the discipline.

The foregoing realities should not dissuade anyone from becoming familiar with the tenets of the science relevant to his or her professional (or personal) interest. Instead, the realities of the world projected by the natural sciences should cause one to approach an ecological system with considerable respect and a realization of the possibility that ignorant or uninformed intervention may do more damage than the harm perceived to be in need of human correction.

For these reasons, it is essential that the natural scientist who is also an environmental professional be scrupulous in the practice and the art of science. Value judgments need to be made overt, first and foremost to the scientist himself or herself. Thereupon, the environmental scientist should be prepared to honor the injunction of the Federation of American Scientists to avoid dogmatic claims, to admit and correct errors, and to reason with those in disagreement, especially in public debate.

Chapter 2

Scope of Ecology

Ecologists speak of the web of life in recognition of the amazing array of relationships and interconnections through which organisms interact with their environments, a concept as narrow as the surroundings of a single cell and as grand as the biosphere.

An organism's environment even includes other organisms of its own and a wide variety of other species. The relatively new science of ecology represents the rules of the house, the name coming from the Greek phrase *oikos* ("house") *logos* ("governing rules"). The rules are deceptively simple, but they must be heeded by environmental professionals as well as by ecologists, even as they must be followed by biological systems. An overview of ecology can be summed up in a few sentences. An ecosystem must have resources and conditions able to support its residents through all the phases of their respective lives.

Each resident is obliged to earn its keep by performing some function useful to the ecosystem. All the work of the ecosystem must be accomplished; that is, all the functional niches are to be filled. A candidate for entry into an ecosystem's biota must be accommodated to and by its fellow organisms and the abiotic environment. In the hypothetical closed ecosystem, the residents cannot leave, and new individuals or populations cannot enter to rejuvenate the community or to replace its members.

In many ways, scientific efforts are like any other human endeavor-only more so. It is in the nature of both to be reductionist, if only in selfdefense. The only way to take on the immensity of the natural world, building out of the paradoxical interweaving patterns from a surprisingly few straightforward foundations, is to pick apart a few of the patterns at a time.

The biota of earth and associated ecological principles and concepts are particularly complex. Simplification is an essential prerequisite on the way to comprehension. A contemporary research ecologist has bemoaned the complications experienced in attempting to work with ecological relationships. In order to begin sorting them out, the ecologist must engage in the Herculean strategy of putting some of the components under a rock while meeting with the others. Unfortunately, simplification holds hazards for the unwary "by chopping biological systems into small pieces, scientists may be destroying basic selfcontrolling and buffering mechanisms and actually making some aspect of them *more* difficult rather than easier to understand,"

The wise ecologist acknowledges the role the laboratory can play in sorting out ecological realities. But there are impediments here, too. Among the most insidious is the nature of the species most suitable for study in the laboratory. In a word, they tend to be "weeds." As such, these species are more adaptable than most. "They lack the highly specific requirements that imply coevolution in organized communities."

In their current state, the natural sciences are far removed from ordinary intellectual endeavor. While not as remote as the physics of subatomic particles, ecology has become distanced from the comfortable ruminations of the worlds of the naturalist, in the field or in an armchair or mesmerized before the television set.

We have all lost something as a result. There was once a time (in recent centuries) when an educated person could grasp at least the outline of the whole of human knowledge and enjoy the option at least of proceeding in more personal realms upon that foundation.

But since those not-so-remote times, science has outpaced us. Theories "are never properly digested or simplified so that ordinary people can understand them." Even the specialist can hope only for a proper grasp of one small piece of the whole. As a result, we are bewildered by the world. But we yearn for comprehension and seriously inquire of the nature of the universe and why it is the way it is.

In each of the last few decades, these yearnings have taken on new practicality as the environmental movement has become national policy codified into law. For all their remote complexity, ecology and the environmental sciences are the essential foundation for the decades of the environment—past, present, and future.

ECOLOGY AND ENVIRONMENTAL SCIENCES

Ecology and the other environmental sciences are not to be divorced from each other. In a very real sense, all the natural sciences are to some degree environmental in their application or their implications. Ecology is a legitimate contender for identification as the culmination of all the sciences.

At one level, it is important to understand that physical and chemical conditions are the essence of ecology and environmental sciences. Each abiotic factor, physical or chemical, exhibits a minimum, optimum, and maximum for every biological system, from the subcellular to the ecological level. If a factor occurs below the minimum, the biological entity will be unable to function. If one occurs above the maximum, the entity will be "poisoned."

Frequently, the optimum, the quantity best serving the entity, is near the maximum tolerable. As will be discussed in greater detail in the context of transitions between the purely chemical and physical components of ecosystems and the biota, mixtures of conditions are actually affecting biological systems in rapidly altering quantitative terms, which can be difficult to follow. Thus, minima, optima, and maxima of individual components are not so readily detectable in real ecosystems as they are in isolated situations. Fortunately, the Theory of Integrative Levels serves ecology and the environmental sciences.

Within the ecosystem and its community of populations,

each individual organism has the innate capacity to follow the pattern that is the life history of the species, but it will exhibit Behaviour patterns that, although unique, are within the constraints posed by the species and will adapt to circumstances in ways that range from momentary physiological accommodations to Behaviour responses.

Similarly, the species itself has a life history, exhibits Behaviour unique to it, and can adapt to changes in the environment. So, too, do ecosystems, when studied over relatively long intervals, reveal a life history consisting more of adaptation and change than of *Behaviour* as that term is usually understood.

While ecosystems are structural and functional units represented by relatively stable biological communities, their boundaries are to a certain extent artificial and arbitrarily established. Aquatic ecosystems, even oceans, are inextricably linked to terrestrial systems, yet they and portions of them can legitimately be isolated for analysis without compromising the quality of the results attributed to the system as such.

Similarly, although terrestrial ecosystems can be isolated and identified by several classification systems, in large part associated with a combination of latitude and altitude, there are no totally independent ecosystems. Moreover, there are "specialists" among organisms that thrive in habitats between ecosystems. There are, in short, not just overlaps but microcosms within microcosms within (and between) ecosystems. The complex of ecosystems forms the planetary biosphere. Each can influence any or all of the others. Despite the internal reality and integrity, these systems are *inextricably* connected. The environmental professional who loses sight of that reality will almost certainly live to regret the oversight.

Since everything from the physics of fundamental particles to the history of the universe is connected and is at least arguably focused in the ecosystem, it is invaluable to pause here for reflection on two extraordinarily profound aspects of the scientific endeavor: cosmology and the relationship between scientific order and the relative newcomer to science, chaos theory.

COSMOLOGY: FROM THE INFINITESIMAL TO THE UNIVERSAL

There was a time, long before the time when legitimate ecologists could be ensconced in armchairs, when the cosmos could be expressed in five Aristotelian terms. The whole of the natural world emerged from four basic "elements"—earth, air, fire, and water—combined with two forces-gravity and levity. This was a time when matter was deemed continuous. It could be subdivided into ever-smaller bits without limit. There would never be a point at which the material of the world was no longer divisible.

Later, atoms were identified and held to be indivisible. Centuries later, the precise relationships among atoms and molecules were detected. As recently as 1905, Einstein observed that brownian motion—the constant joustling of protoplasm and other suspensions viewed with a light microscope—is actually the result of collisions between atoms and particles large enough to be visible. In another few years, the internal structure of atoms-the complexes of protons and neutrons surrounded by clouds of electrons-was discerned. Now we have discovered protons and neutrons to be precisely divisible into oddities known as quarks bearing features physicists whimsically refer to as "flavors" and "colors."

All of this is quite beyond ordinary powers of comprehension. Earth, air, fire, and water are far more accessible. At the same time, gravity, if not levity, has transformed into four forces, including some associated with subnuclear particles. In addition to the familiar, but no less mysterious, gravitational force, there is the electromagnetic force as well as the weak and strong nuclear forces. While gravity may be familiar, it is not as well understood as might be suspected. Although it is both universal and consistently a force of attraction, gravity is the weakest force and is mediated by the graviton, a virtual particle without mass. It is this infinitesimal world of vanishingly small particles and associated forces to which cosmologists turn for explanations of the origins and destiny of the universe.

In a very real sense, energy and matter are wholly interchangeable, if not actually identical. In a comparably real sense, ecological sciences are derivatives of both extremes in the scientific scale. The biosphere is one expression of both fundamental physics and cosmological events. But this realization does nothing to advance precision in predicting events from causes, as was once a hope, if not an expectation, among scientists.

ORDER AND CHAOS

Because of the emerging simplicity and order of the natural world, a mission was advanced for at least a moment of human history. When physical phenomena were extended into the biological and then into the psychological and social realms, a kind of giddy optimism followed.

Among scientists and observers of science, there were those who sincerely believed every event explicable "in terms of antecedent events in causal chains and networks, characterizable in terms of universal laws which make no reference to the causal efficaciousness of future events or higher levels of organization." In the early nineteenth century, arguments were seriously advancing the concept of a complete determinism. Scientific laws would allow predictions of "everything that would happen in the universe, if only we knew the complete state of the universe at [any] one time." Since "everything" included the vagaries of human Behaviour, it certainly included more scientific environmental processes.

Promise or threat, such scientific predetermination has now been dismissed. Scientists and those observers of science are coming to accept that life, at the very least, is "not meant to be free from contradiction or ambiguity."

It is only in this century that determinism has lost credibility in the face of advancing science. That level of order gave way to the Heisenberg Uncertainty, which has revealed that one cannot accurately know both the position and speed of subatomic particles simultaneously. The new science of quantum mechanics arises empirically from mathematical manipulation of the components of the Heisenberg Uncertainty to predict not a single, definite result but the array of possible outcomes and the likelihood of each. One of the resultant insights crucially important in ecology is that light, while composed of waves, behaves as if particulate. Light energy is emitted and absorbed only in the packets known as quanta.

Primarily a practical branch of physics, quantum mechanics explains phenomena as wide reaching as properties of collapsed stars, conduction of electricity, structure of atoms and their nuclei, chemical bonding, and the mechanical and thermal properties of solids.

In more recent days, scientists have come to recognize that while chaos seemingly arises from the orderliness of scientific law, order is the mark of chaos. A new kind of highly mathematical science has reached the headlines and fiction, not to mention volumes intended for coffee tables. In the end, the important conclusion is that life consists of orderly and lawful Behaviour on the part of matter.

Its tendency is not exclusively to proceed from order to disorder; living matter is based on keeping up existing order through metabolism. Not even large biological molecules like DNA escape or invalidate physics, chemistry, or the laws of probability. But none of them can adequately explain life or predict the direction of cell, species, community, or ecosystem. The reason is embedded in chaos.

A WORLD OF ECOLOGIES

Ecology could hardly remain untouched by novel scientific insights, on the one hand, and the political implications of an evolving environmental movement, on the other. There can be few in this country who do not subscribe to the concept of a healthy, diverse ecosystem in a state of dynamic balance. Within such an ecosystem, there are cycles and chains. Food chains connecting animal to plant direct energy outward, whereas death and decay restore energy to the earth. Whenever a change occurs anywhere within the circuit, the remaining parts must adjust to restore the balance of the whole. Such a description of ecological balance might be uttered by either a practicing ecologist or an environmental professional. It could also be a political statement, for ecology has become a matter of policy and politics.This description might also be a reflection of the spiritual realm and, as such, could arise from any number of traditions, from the Judeo-Christian to the Native American to the Asian to the primal, extant as well as extinct. Ecology is no longer the sole province of the scientist working in aloof isolation.

DEEP ECOLOGY

Today, three ecologies—scientific, political, and spiritual—are merged in what is called deep ecology. Yet the proponent of deep ecology need not be a scientist or even any one of the three kinds of ecologist. *Anyone* can be a deep ecologist. In fact, scientific ecology, especially as practiced by environmental professionals, is viewed by some as the antithesis of deep ecology, which is more a discipline of philosophy expressed in poetry than any other form of science.

The mission of the deep ecologist is to cultivate individual but widespread ecological consciousness with an insight into the ubiquitous connections of the world around us. Deep ecology is holistic. Its message is one of appreciation, receptiveness, and trust. Whether one reacts to deep ecology with empathy or impatience, its tenets proffer much for contemplation by the environmental professional and the interested nonscientist:

- All life has intrinsic value.
- The richness and diversity of life have value.
- Human life is privileged only to the extent of satisfying vital needs.
- The relationship of humans to the natural world endangers life's richness and diversity.
- Maintenance of life's richness and diversity mandates a decrease in human population.
- Changes are needed to accommodate cultural diversity affecting basic economic, technological, and ideological components.
- Ecologically sensitive ("green") societies value quality of human life over quantity of human life.

How these tenets fit into the ecology of populations, especially of humans, and into the matters of the place of the biota in the strategy of ecosystems will be explored throughout this book, with special reference to the concluding chapters. Whatever the scientific realities of the ecology of the human species, deep ecologists are obligated to promote the needed sociocultural changes. Surely, even these philosopher-poets will be more effective if informed by ecological realities from the scientific realm.

GAIA HYPOTHESIS

SCIENCE OR METAPHYSICS

Ecologists, deep, scientific, or political, have come to recognize this planet's biosphere as a distinct level of organization arising from the interconnections among all the ecosystems. If an ecosystem is a biological system, logic is not strained through the perception of the biosphere as the highest of all biological systems. There are those who would take one more step, perhaps moving from science into metaphysics in doing so. Is the biosphere itself actually a biological entity—a living organism?

This is the position advanced by the Gaia Hypothesis. There are valuable insights to be taken from Gaia, whether one is a believer or a skeptic content with stopping short of Gaia with the biosphere as nothing more (or less) than the highest level of organization.

J. Lovelock propounds "that the entire range of living matter on Earth, from whales to viruses, and from oaks to algae, [can] be regarded as constituting a single living entity, capable of manipulating the Earth's atmosphere to suit its overall needs and endowed with faculties and powers far beyond those of its constituent parts."

An intriguing "branch" of the Gaia Hypothesis asserts bacteria to be the primary cybernetic mechanism at Gaia's command: "The environment is so interwoven with bacteria, and their influence is so pervasive, that there is no really convincing way to say this is where life ends and this is where the inorganic realm of nonlife begins." According to the variant of evolutionary theory and of the Gaia Hypothesis, bacteria go beyond filling every known habitat on the earth's surface to maintain all life forms and to accelerate all biological cycles. In short, bacteria represent a complete planetary regulatory system.

Biosphere or Gaia? Is the dominant taxon of earth the kingdom of Monera (bacteria) or *Homo sapiens sapiens* or some organism ecologists have overlooked? Are there practical differences among these distinctions? The individual reader must decide. Armed with the definition of life, the theory of Integrative Levels, and an awareness of how science works and how the several ecologies fit into human endeavor, the reader will be provided in the chapters to follow with principles and concepts to inform a personal conclusion based in ecology.

Whether one believes in the Gaia Hypothesis or in a variant perspective on evolutionary theory or simply in a functioning biosphere, there is an ecological truth for acceptance by all: "Consisting of life, the environment is continually regulated by life, for life."

THE GENERIC ECOSYSTEM

If, as was asserted in the opening chapters, an ecosystem can be seen as a black box, representing either one of the highest levels of scientific organization or a kind of biological system, what are its internal black boxes? The ecosystem's complex of inseparable interrelationships is organized in both structure and function. The two can be characterized most generally by the flow of energy through the system and the cycling of materials within it.

In short, an ecosystem is made up of habitats, locations in the three spatial dimensions occupied by populations of organisms interacting in a community. If one were to construct an ecosystem, the natural point of departure is the setting composed of habitats, the physical and chemical abiota to which organisms come.

The organisms—biota—are organized in relationships

among individuals composing populations and the community characteristic of the setting. The resultant whole is the ecosystem. All the relationships of this biological system are a causation network arising from interactions that are obligatory and interdependent. Close examination of an ecosystem reveals that it is actually made up of multiple smaller units, or microcosms, in which only a few organisms are directly participating, although they somehow contribute to the structure and function of the larger whole. This larger level of analysis is the holistic approach to ecology.

The distinctive biological system of the ecosystem affords each unit at this level of organization its identity. Each species present constitutes a population distributed in its particular habitat and serving the whole in a niche, a structural and functional role of which habitat is only one part.

STRUCTURE AND FUNCTION

HOLISTIC ANALYSIS

The contents of the black box known as an ecosystem are distributed in the abiotic and biotic realms. The former realm is characterized by low energy and inorganic compounds; the latter, by high energy and organic compounds. Energy flows through the system in patterns that follow the overall pattern describing an ecosystem as a biological unit. With the flow of energy, materials cycle between the two realms and among the components of the biota.

The major functional component of the ecosystem lies in the interaction among autotrophic and heterotrophic organisms. The principal features of this interaction are trophic structure, material cycles, and biotic diversity. Both the degree of and the participants in biotic diversity are used not only to identify but also to describe the ecosystem and its condition. Any deviations from anticipated diversity may be an indication of the health of the system or its component parts.

Working with the geological substrate, autotrophs fix light energy and utilize inorganics and some simple organics to synthesize and accumulate organics of greater complexity. The stuff of physics and chemistry is converted into living, functioning organisms. Heterotrophs utilize relatively complex organics constituting plant life for rearrangement as the large organic molecules serving their own life processes. Other heterotrophs decompose these complex materials, returning the basic materials to the substrate as increasingly simple organic and inorganic molecules.

This array of functions is intimately associated with the structural component of the ecosystem. Structure is distributed among abiotic and biotic features. Abiotic structure is physical and chemical and found among inorganic and organic substances as well as in the climate regime, characteristic interactions of light, temperature, moisture, and wind. Biotic structure is distributed among producers (autotrophic), macroconsumers (heterotrophic phagotrophs) and microconsumers (heterotrophic saprotrophs).

Neither the populations nor the community of organisms is a mere aggregation. Based on the activities and interactions of its individual members, each population is itself a unit exhibiting structure and function. The interactions among populations in an ecosystem give rise to the structure and function of its distinctive community. The whole depends on the parts—abiotic as well as biotic.

Stability, known as homeostasis, can be discerned over different time scales. Homeostasis is maintained by cybernetic mechanisms, a variety of control devices operating at every level from subcellular molecular biology to changes in the ecosystem at large. Thus, the components of an ecosystem will be shifting subtly as environmental conditions constantly alter over time measured in units smaller than seconds. Larger alterations over longer intervals may give rise to adaptations, which may either be retained or lost as new ones assert themselves.

Discernible stability notwithstanding, every ecosystem is undergoing development and, over longer intervals of time, evolution. Adaptation can be distinguished from evolution in the former's smaller time frame and temporary nature requiring no alteration in genetic structure. Adaptations are reversible and range in units of time from fractions of seconds to the lifetime of the affected biological system. By definition, evolution occurs over several generations of a given biological system and proceeds through inheritable genetic alteration.

Behind the ecosystem as a biological unit is the entire array of sciences we distribute among physics, chemistry, and biology. The biological nature of the organisms in an ecosystem determines their roles. That nature can be understood through the biochemical reactions occurring at the subcellular level and expressed ultimately at the organismal level as each individual participates in the interactions of its population and community.

The energy that is driving the overall process comes from the sun and participates in the chemical reactions occurring in the cells of the biota. Put most simplistically, energy enters the black box from which emerge the products of respiration and energy in the form of heat. It is instructive to so view the generic ecosystem. In contrast, most of the material in a well-established ecosystem is available within its boundaries.

Sunlight enters along with water and carbon dioxide and a few other inorganic substances. Molecular oxygen, water, and heat exit. In a sense, this exchange can be viewed as the whole of an ecosystem, but it is hardly a satisfactory description. For increasingly satisfactory descriptions, it is necessary to explore the contents of this black box and all the black boxes of which it is composed.

Each of these analytical approaches to empirical ecology can be followed at any level, from the subatomic to the biosphere itself. Each level of biological organization from the single or individual cell to the biosphere has a life history and what can be called Behaviour. Each can adapt, and each can undergo evolutionary processes.

REDUCTIONIST ANALYSIS

Analysis of structure and function becomes possible by taking a reductionist approach. Every ecosystem is made up of essentially the same components, each in patterns unique to it. There are energy circuits and their corresponding food chains and webs. There are cycles of matter in the form of biogeochemical cycles that suffuse all the components of both realms. There are patterns of diversity that can, and must, be described in terms of three-dimensional space and in terms of time as the fourth dimension. In different language, there is an economics of an ecosystem: Resources are allocated, cost-benefit ratios operate, and optimization occurs.

The resources are energy and materials, the cost-benefit ratios are based on energy distribution and such considerations as competition and predation, and optimization results when organisms occupy niches.

Energy Flow

Much of what makes an ecosystem a discernible unit can actually be explained in terms of the First and Second Laws of Thermodynamics. The First Law of Thermodynamics states that neither energy nor matter can be created or destroyed. Each, however, can be transformed from one form to another. Energy, for example, is generally characterized as potential or kinetic. Energy, described as the capacity to do work, is considered potential when it is poised to do work and kinetic when it is in the process of doing work. Familiar forms of potential energy are a rock balanced at the edge of a precipice and water at the top of a gradient. Kinetic energy is represented by the actual fall of the rock or the flow of water.

Less familiar forms of potential energy are in a relaxed muscle, the internal bonds holding molecules together, and a nerve synapse. The corresponding kinetic energy of each is the contraction of a muscle to do mechanical work, the breaking of the chemical bond to release the contained energy, and the electrical charge of the activated nerve cell. Each contributes to the functioning of organisms, and organisms are constantly interacting in and with their environment.

While all the energy of ecosystems comes from the sun, the forms that solar radiation takes are greatly varied. It can become biomass or wind or tide or heat. Energy in ecosystems is, then, a vivid demonstration that energy and matter are interchangeable.

Energy is not created in an ecosystem but biologically fixed as it enters from outside—that is, assimilated through physical and chemical processes into a form available to life. That form is matter, the material of which biomass is composed. Similarly, matter is not created in an ecosystem but undergoes a series of alterations in chemical composition.

For every reaction of life, there is some loss of energy. This is an inescapable reality imposed by the Second Law of Thermodynamics, which can be stated in a number of different qualitative ways—accepting without further discussion the sophisticated mathematics describing the principle in precise quantitative detail.

The Second Law holds that, left to itself, every system will dissipate into complete randomness—total entropy. Although chemical reactions occur, none is 100% efficient, some energy is lost in the process, usually as heat, measured as entropy. Both structure and function are constantly working against entropy. In a universe proceeding toward entropy, life is often viewed as the extreme of organization consisting of negative entropy.

The processes of energy flow through an ecosystem can then be summarized as the entry of radiant energy from the sun, some of which is reflected outward through the atmosphere, into the solar system, the galaxy, and the universe beyond, where it is lost to the planetary processes of our solar system. Some of that energy, however, becomes a part of the abiotic realm. It will warm the atmosphere and the earth to temperatures allowing life processes to occur. It will take the form of wind, which contributes to climate on a global and local scale. Some will become a part of the global hydrogeological cycle, which is an important source of both energy and material in ecosystems.

A fraction of the sun's radiant energy, specifically that occurring near the midpoint of the visible range, is captured by the biota. It is absorbed and assimilated by green plants and other autotrophic organisms. By the process of photosynthesis, the sun's light energy is converted into plant biomass. In a series of steps, that biomass may be consumed by populations of animals. When these animals excrete wastes, the waste products become nutrients to other organisms from animals to fungi and bacteria. These organisms return the materials of the biomass to the physical realm, where they can reenter the processes initiated by photosynthesis.

Material Cycling

It should have become obvious rather quickly that it is difficult to isolate one part of an ecosystem from another. While energy flow can be isolated for close analysis, the relationship to biochemistry cannot be neglected. It is chemistry, then, that describes the countless steps by which assimilated energy is distributed throughout the ecosystem by the biota. Green plants and other autotrophs initially assimilate the energy and convert it into their own biomass and use it to conduct the processes of their lives.

Homeostasis

Homeostasis is a tendency to resist change and remain in a state of equilibrium through processes of self-maintenance and self-regulation. These are the processes of cybernetics, the study of how biological systems maintain the status quo or maintain their integrity at a different level of homeostasis. Equilibria are maintained by positive and negative feedback.

Adaptation and Evolution

Despite the tendency of biological systems to restore altered order through cybernetic processes reestablishing homeostasis, no biological system, including ecosystems, is in any sense static.

As a result, no description of an ecosystem or of its living components can be expected to remain accurate indefinitely. Adaptation is a constant process leading to evolution or extinction whenever perturbation exceeds the cybernetic capacities of populations.

A description of an ecosystem is best acknowledged as a snapshot. It will be as accurate a representation of the real thing as the camera and film of ecological analysis allow. But tomorrow's snapshot is likely to be quite different from today's, whether the equipment is the same or an improvement.

Furthermore, although ecosystems are more than a construct to aid in scientific or other analyses, their boundaries are not so clear-cut as scientists and other observers make them. It is obvious that energy comes from outside the ecosystem in the form of light, heat, wind, and physical power of water. Energy also leaves the ecosystem. Similarly, material and even organisms may come from beyond both the convenient and the real boundaries of any given ecosystem or somehow be carried beyond them.

It is helpful to consider these factors in two kinds of ecosystems—terrestrial and aquatic. The ecologist must be prepared to examine both the minutia of the ecosystem and its larger setting.

Chapter 3

Global Ecosystems

Every ecosystem depends on a mixture of abiotic and biotic components, but it is always greater than the sum of those parts. In examining global ecosystems, there are any number of points of entry. In keeping with moving from the whole to the abiota to the biota, there is some logic in moving from habitat to community and finally to the ecosystem as the resultant whole. This, after all, is the strategy of the individual ecosystem and of the biosphere.

Habitat is where an organism lives. If habitat is an organism's address, niche is its line of work and actual day-to-day employment. Communities are the living connections by which habitats are linked to an ecosystem. The ecosystem emerges as the functional biological system consisting of specific structural components. Aquatic or terrestrial, each ecosystem will follow the generic pattern provided in the last chapter but with variations unique to it and its location.

AQUATIC ECOLOGY

THE HABITATS

From clear mountain stream to swamp, from brook to pond to sluggish river to freshwater inland sea, from shoreline to abyssal depths, from polar cold to equatorial tropics, from centimeters (or inches) to kilometers (or miles) in depth—aquatic systems represent a variety of habitats at least as diverse as their terrestrial counterparts. On the basis of salinity, three aquatic environments are recognized: freshwater, marine, estuarine.

Freshwater habitats are broadly classified as standing water (lentic) or running water (lotic). In accordance with the general rule, sharp demarcations from one habitat to another seldom occur. Moreover, a detailed nature of each will depend greatly on surrounding ecosystems, terrestrial as well as aquatic. Temperature, transparency, and current are the primary features of freshwater habitats. All three impose physical and chemical influences on the biota. Temperature ranges are relatively limited.

One of the amazing accidents or coincidences of the biosphere is that while water freezes at 0°C (32°F), it is at its densest point at 4°C. Water is lighter above and below that temperature. The ecologically significant result is that ice floats. The significance to the biota is that large bodies of standing water often remain liquid throughout the cold seasons, allowing living processes to proceed.

Another ecologically important feature associated with water's unique physicochemical properties is the turnover phenomenon. In spring and in fall, the surface and deep waters of large lakes exchange places—turn over. The result is a thorough redistribution of nutrients. Primary producers, of course, depend on the penetration of light into the water column. Transparency of water is also a physicochemical property, but it is greatly influenced by the presence of suspended particles, chemical and living. The patterns of current, too, have physical and chemical implications to which freshwater organisms have adapted through evolutionary processes and must adapt from moment to moment and through seasonal changes.

Additional features composing freshwater habitats are concentrations of dissolved oxygen and carbon dioxide, essential to respiration and photosynthesis and important to concentrations of positive and negative ions (pH). This pH and related buffering capacity of the water are at once mediated by and crucial to the chemistry of surface waters.

Chemical identity, mobility, and biological availability depend in part on whether pH is acidic, basic, or neutral. Most notable among these chemical substances are the biogenic salts, particularly nitrates and phosphates. Strangely enough, the oceans—in which life originated and still a major source of material for terrestrial populations—are accurately described as semideserts. For lack of energy and material, the open ocean is unable to support dense communities.

Still, as an array of habitats, the earth's sea is notable for its size (70% of the planetary surface), its depth, and its continuity. Barriers to movement, abiotic or biotic, do exist as a result of temperature regimes, differing salinity, and depth. Each organism will be subject to a specific tolerance range for combinations of these factors.

In addition to light and temperature, a major source of energy in marine habitats comes in the form of currents, waves, and tides. Comparable to terrestrial climatology and resultant meteorology, continuous circulation of ocean waters affects marine habitats physically, chemically, and biologically.

Literally standing between freshwater and marine habitats, and heavily influenced by both, are estuarine habitats. These are the aquatic ecotones, transitions between freshwater and marine ecosystems.

As such, estuarine habitats are more than transitional; they are in some ways characteristically unique. In addition to the mingled influence of flow, currents, and tides, there are the rhythmic alterations in salinity. Although these habitats are rich in nutrients, species diversity is kept low because of the demands for wide tolerances.

Not surprisingly, there is a wide variety among estuarine habitats. Not only do they alter with the terrain and bodies of water forming them; estuaries will also be influenced heavily by their location on the globe corresponding to, and affected by, the surrounding terrestrial ecosystems.

As is always the case, care must be taken in drawing boundaries around aquatic ecosystems. It is well established that the oceans not only compose a major component of hydrogeological cycles but also bear a heavy influence on global climate and meteorological patterns. The oceans also have a substantial place in large-scale material cycling, with special reference to carbon.

THE COMMUNITIES

Freshwater habitats support the same kinds of food webs readily identified in terrestrial communities. There are the autotrophic producers and the heterotrophs, phagotrophic macroconsumers and saprotrophic decomposers, and microconsumers. The distribution of communities covers all three dimensions.

Near the surface are communities of phytoplankton and the zooplankton, the latter only marginally more motile than the former. Throughout the body of water are the swimmers—large insects, amphibians, and fish. Some of these organisms are quite capable of moving to the surface or to the bottom substrate to feed or to proceed through the stages of their life histories. Where currents and transparency allow, certain organisms attach to the underlying substrates along the edges of the water body and into the benthos, the substrate beneath the water column. Communities expand to include plants and animals that attach to these attached organisms.

The benthos is rich in life, forming a diverse community in and on the solids. In each ecosystem, the actual taxons composing these communities are determined by the composite of abiotic features characterizing each body of water all the way from its general geological formation to the microenvironmental scale.

Marine biota also include plankton, swimmers, some attached plants, and associated plants and animals and benthic communities. Both the taxons and their representatives may be unique to the marine environments. For example, sponges and echinoderms, relatively unimportant in fresh-water communities, are highly significant to marine ecology.

Higher plants are rare, and insects are replaced by crustaceans. In marine habitats, large mammals travel huge distances in groups and are among both primary and secondary consumers. Bacteria, algae, crustaceans, and fish, however, are dominant in both freshwater and marine environments. The representative communities of the two are distinct and, as always, characteristic. Moreover, communities filling the respective habitats of freshwater and marine ecosystems are quite characteristic. It is worth noting that some of the species present possess elaborate life cycles, some of which can be highly mysterious even to the most observant of researchers. This is particularly the case among marine organisms, which may fill niches in widely disparate habitats from one stage to the next. Along the shorelines and estuaries, marine and aquatic birds are a part of the intricate food webs. Wetlands and freshwater formations support their own communities, permanent and transient.

Many terrestrial mammals pass much of their time in and along such bodies of water. Aquatic fowl and the raptors are semipermanent residents. Finally, migrating flocks of fowl and songbirds depend on aquatic habitats to support them in their extraordinary intercontinental flights. But estuaries may be most significant ecologically as nurseries serving an astonishing array of populations and, of course, the sometimes-distant communities in which they participate in subsequent stages of their life cycles.

THE ECOSYSTEMS

Every brook, pond, lake, stream, river, inland sea, estuary, swamp, bog, or ocean, each with its respective community, can be legitimately identified as an ecosystem. On one scale, it may be foolish to take on ecological analysis of the whole of the aquatic portion of the planet, but it is unwise to isolate any aquatic habitat from the larger environment.

Numerous ecosystems, each marked with its own community, are generally recognized by aquatic and marine ecologists. The existence of some, if not their ecological detail, is quite apparent to the most casual of observers. In freshwater systems, there are shallow (littoral) zones, open water (limnetic zones), and deep waters (profundal zones). In flowing waters, there are distinct rapids and pools, as well as more intricate or subtler formations. Each combination of distinctive habitat and corresponding community can be approached as an ecosystem, whether relatively open or relatively closed. Not to be disregarded is the complete identity of the underlying substrate, which is particularly important in the matter of potential perturbation attributable to chemical substances as either nutrients or toxins.

Lakes are large enough to exhibit temperature stratification, which will exert a major effect on the distribution of both nutrients and organisms and which will alter energy availability and demands. Ecosystems will reflect those dynamic influences in the detail of ecological structure and function. Marine ecosystems exhibit similar zonation and stratification, ranging from shallow intertidal zones to the open ocean where there are the same kinds of geological structural formations as characterize terrestrial geology. In terms of productivity, the estuarine ecosystems outperform both marine and freshwater counterparts. High productivity is attributed to the energy subsidies occurring only in these environments. Moreover, estuarine ecosystems are nutrient traps. Thus, they are amply endowed to serve communities rich in numbers if not diversity.

From the human perspective, estuarine ecosystems not only are a bountiful source of food but also are capable of the ecological (re)cycling of material wastes generated by human life and industry. The designers of Biosphere 2 took this function into consideration in seeking methods to cleanse human wastes from this ambitious but severely limited artificial ecosystem.

It is more in keeping with ecological realities to be prepared to work from microcosm to an ever-expanding definition of the ecosystem in order to discern just what is entering and leaving the particular black box of immediate interest. Individual communities are functioning in defined ecosystems within each larger body of water. Similarly, it is essential to look beyond the body of water to larger watersheds or terrestrial ecosystems influencing and influenced by that environment and its biota.

THE TERRESTRIAL ENVIRONMENT

The terrestrial ecosystem is built quite literally from the

ground up. Terrestrial substrates vary widely from cooled lava or volcanic ash, bare rock, sand, and rich or poor soil. Whatever their physical and chemical nature, global substrates attract characteristic plants and microorganisms.

These are the primary producers, microbial autotrophs, and the saprotrophs. Each association among these organisms is in turn associated with corresponding herbivores. Insects, rodents, birds, hoofed mammals—their identities depend on the specific plant taxons present. Communities within communities will act together, some in obligate pairings or more elaborate combinations, others in more flexible, temporal associations.

The community of carnivores will include many of the same orders and higher taxons as the herbivore community, composed of the same or different genera and species. An ecologist or a taxonomist can identify the location by describing the ecosystem and identifying the species making up its community.

It is possible to trace characteristic changes in ecosystems with distance in a line perpendicular to a shore front, lakeshore, or band of flowing water. The changing habitats support different plants, which bring different associations among the remainder of the community.

Height becomes increasingly important. Low-growing plants—lichens, mosses, and grasses-give way to shrubs, bushes, and small trees, then to taller, more luxuriant, and older trees. Characteristic animal associations occur from soil to treetop. Some are permanent; others come and go in diurnal or seasonal patterns; still others are transient.

Terrestrial ecosystems depend primarily on climate and substrate. Almost all other important abiotic factors are related to that combination and to the modifications effected by the biota. For this reason, the classification of terrestrial ecosystems is important to the ecologist and to the environmental professional.

Whenever one knows what should be present in a given location, it is possible to determine whether some interference with the ecosystem is occurring.

CLASSIFICATION OF TERRESTRIAL ECOSYSTEMS

Methods of Classification

Because they tend to be relatively fixed in place, terrestrial ecosystems lend themselves to classification. In addition to opening the way to understanding structure, function, and strategy of an ecosystem, classification points the way to determining the condition of a specific ecosystem.

There are a variety of ways in which ecosystems have been classified, and each reflects an internal ecological logic. Zoogeography provides a classification identifying biogeochemical regions. Temperature regimes support life zones with characteristic plants and animals. Ecological associations at the species and subspecies levels give rise to biotic provinces, with an emphasis on the ranges and centers of distribution of animals. Here, plants lose no importance to the ecosystem, but boundaries are drawn from physiographic barriers rather than from types of vegetation.

Plant formations in their turn provide for biomes, which include associated animals, since animal life depends, after all, on the plant base. The biome takes into account the distinct community of plants and animals that combine in the climax community, as well as the sere proceeding to that relatively stable and long-lived ecosystem. This classification system also reflects the crucial and interacting roles of latitude and altitude. The biome is isolated through a combination of mean annual temperature and mean annual precipitation, climate tempered by local meteorology.

While each classification has its obvious advantages tempered with disadvantages, depending on the needs of researcher or observer, each should complement, if not actually confirm, the conclusions afforded by the others.

Still, there is one classification system that seems the most promising for environmental professionals: the Holdridge Life Zone System. Based on three predominant interacting global features—latitude, altitude, and moisture regime—which are very closely related to climate, the Holdridge life zones should represent or predict climax communities throughout the globe.

Each life zone, defined by climate, will support a specific kind of community, the specific taxons present further determined by local conditions.

The Holdridge system can be represented by a triangle containing sixsided figures, each representing a zone. Across the base of the triangle are humidity provinces from the arid to the humid. Along each side, temperatures decrease from base to tip, on the one side as latitude moves from equator to pole and on the other with increasing altitude. The two are comparable in effect on community type in that both distance from the equator and altitude affect temperature, expressed as biotemperature, mathematically manipulated to represent the extremes organisms tolerate.

Thus, tropical latitudes are comparable to lower montane altitudes, and subpolar to alpine. The internal hexagons each represent a different kind of community. At the arid extreme, deserts occur at every latitude or altitude until the most extreme dry tundra occurs. In temperate latitudes or altitudes, moisture increases allow deserts to be replaced with desert scrub, steppe, moist forest, wet forest, and finally rain forest.

A newer classification system, identifying ecoregions, is a synthesis going beyond climate type to include actual associations of vegetation and soil type. Underlying it, and an implication behind the other systems as well, is that uniformity of climate and soil should lead to similar vegetation. Disparities, then, should always reflect locally unique conditions, an early seral stage, or perturbation.

IMPLICATIONS FOR ENVIRONMENTAL PROFESSIONALS

The foregoing has been no more than the sketchiest of global ecosystems and their significance to ecologist and environmental professional. Little justice has been done the similarities, distinctions, diversity, and richness of global ecosystems. Nevertheless, the connections significant to the practice of any environmental professional are extracted from the full array of those considerations. The basic structural and functional ecosystem components are to be expected, but the expected biota will be a unique and characteristic community. The initial determinations the environmental professional must make in revealing why an expected community does not occur are the ecological history of the ecosystem, its seral stage, and the nature of the substrate supporting the community.

Only when the naturally occurring features that distinguish the ecosystem have been identified is it prudent to commence a consideration of likely anthropogenic influences amenable to intervention through applied ecology, technology, or the law.

When a perturbation is revealed, its ecological effects subject to the fourth dimension of time should always be carefully analyzed in preparation for introducing any intervention in the name of the environment. Ecological analysis of intervention is especially crucial in the usual circumstances where the expertise of additional disciplines will be drawn into the Programme.

ECOSYSTEMS AND THE LAW

The preceding chapters reveal the extent to which delineation of an ecosystem through sharp, isolating boundaries is an artificial tool facilitating organized studies, empirical, experimental, or regulatory in intent. Delineation necessarily focuses on a limited array of ecological features to avoid the confounding effects arising from phenomenal complexity at this level of organization.

Arising from the literally uncountable features in dynamic operation in nature, complexity is compounded by the contributions of all the levels of organization above and below the ecosystem.

But ecosystems are not entirely artificial constructs for the sake of mere convenience, scientific or political. There is a biological or scientific reality that bears some relationship, however tenuous to the construct. Reality and construct may possess entirely different boundaries and may be more ephemeral than is acceptable to scientists, not to mention politicians and bureaucrats, but the ecosystem is really there.

Specific categories of ecosystems have been recognized

among scientists; counterparts can be found in, or extracted from, statutes, regulations, and associated documents regulators develop in cooperation with scientists. The question that remains for environmental professionals is whether and to what extent operating definitions of ecology and policy are coextensive. Close behind this question is the matter of the nature and extent of definitional divergence as putative ecology becomes law.

FROM ECOLOGY TO LAW

SCIENCE IN THE POLITICOLEGAL SYSTEM

Transitions from natural science to law are never easy. And among the natural sciences, ecology may prove to be the most complex. The first impediment to facile transition is the particular kind of science being incorporated into law and policy. The second impediment relates to the mechanics of incorporation.

H. Bauer would strain "science" through a "knowledge filter" by which the merely subjective and unreliable principles and concepts are eliminated to leave a filtrate of the reliable and objective. Entering the filter are all the human traits marking scientists along with all the rest of us.

These traits incorporate everything from conservatism to wild ideas. According to Bauer, frontier science is the result of the first filtration, which eliminates the nonsense, stupidity, and pseudoscience. From the frontier, scientific thought is filtered into the primary literature with the elimination of bias, error, and outright dishonesty. Through an additional three filters, knowledge proceeds into the secondary literature, textbook science, and ultimately into the textbooks of the future.

All too often, abstract policy and real-world regulation exist in the inadequately evaluated areas prior to filtration and at the frontier of science. Textbook science is often far behind the immediate needs of national politicolegal institutions. As a part of this inadeqate filtration of principles and concepts or their applications to the need at hand, there is the matter of ethics as espoused, respectively, by scientists and by policymakers. Where do the environmental professionals' loyalties properly lie? To the process of ecology or other environmental sciences?

To society or some segment of society? To the employer, public or private? For example, is there an obligation to encourage the sharing of data in the name of ecologically sound policy? Is there any obligation to inform the public at large of the realities of the level of organization when information available to them is skewed?

On another level, how should public research dollars be distributed among ecologists? Should the distribution reflect the call for movement away from enumeration of species and preservation of habitats to elucidation of politically relevant issues such as global climate change or loss of biological diversity? Should the academically unpopular interdisciplinary approaches be fostered?

Actually, these and related questions may be premature in the context of integrating ecology and environmental sciences into policy and law. The more immediate question probably lies in the level of congressional comprehension of environmental principles and concepts as reflected in relevant statutes.

Although natural resources management and related programs are more intimately connected with ecosystems, pollution is the most immediate link between ecology and law at the national level. Pollution can be physical, chemical, or biological and is subject to time considerations. Usually, pollution is described simplistically as too little or too much of one or a few factors, more often chemical than physical or biological. But heat can be a physical pollutant, and infectious or imported organisms, biological pollutants.

Ecologists will discern significant alterations long before an entire population, much less a community or entire ecosystem, is threatened. Subtle changes may occur in the abiota or in the cells of certain organisms. Over time, with continued intrusion or additional pressures, the intensity of the changes will increase and will be manifested in increasingly obvious ways. Some of those alterations will affect natural resources, portions of the abiota or biota of specific, usually economic, interest to humans.

Thus, under the law an anthropogenic agent becomes a pollutant when it affects a natural resource so as to interfere somehow with *human* interests. The significant reference is to human needs or values. At the outset, in accord with constitutional limitations on the legal system, human health comes first, followed by human welfare.

Ecological amenities are a distant third in our politicolegal system. Ecosystems and their components in and of themselves come in dead last. Thus, the interests of a few populations of one species relatively high in the food chain comes first, and subtle alterations within the rest of the biosphere may be ignored until substantial damage has been done.

INTEGRATION OF ECOSYSTEM

To what extent and how well are the ecosystem and related biological systems integrated into federal programs? To a great extent, the answer depends on the avowed mission of the individual Programme. Although conservation and preservation have been evolving since the era of Teddy Roosevelt, the contemporary emphasis is as often pollution control as natural resources management. The benchmark statute of the environmental movement and the one that promised to make the requisite connections is the National Environmental Policy Act (NEPA).

NEPA

This is the statute that opens the door to connecting the national pollution control Programme to environmental protection. A legitimate question here is whether the Congress effectively integrated the concept of *ecosystem* into this crucial enactment. The obvious starting point is in the declarations of policy.

While the word ecosystem never appears, its presence can be felt in NEPA's declaration of purpose to encourage productive and enjoyable harmony between man and his environment; to promote efforts which will prevent or eliminate damage to the environment and biosphere and stimulate the health and welfare of man; to enrich the understanding of ecological systems and natural resources important to the Nation.

In establishing a national environmental policy, Congress continues to emphasize the quality of the *human* environment without express reference to ecosystems. Congress left such detail to those implementing and enforcing NEPA—administrative agencies under the watchful eye of the judiciary.

POLLUTION CONTROL BY ENVIRONMENTAL "MEDIUM"

Air

Close on the heels of NEPA came the first in the array of new or substantially modified federal statutes intended to protect environmental media from pollution through control at the source. It is intended to protect ambient (atmospheric) air through control of industrial emissions. Early versions of the Clean Air Act speak of "mounting dangers to the public health and welfare" without express reference to any potentially affected ecosystems.

Instead, there is reference to the nation's air resources in the context of promoting public health and welfare and the productive capacity of the human population. Even in the context of health and welfare of populations of *Homo sapiens sapiens,* such contemporary issues as acid rain, global warming, and depletion of stratospheric ozone were not among environmental perturbations to be alleviated.

The focus of associated research was more on technology than on ecology or any of the other natural sciences. Nevertheless, at least the inference of an interest in healthful ecosystems is present in the name of public welfare. The matter of protection of stratospheric ozone did enter the Clean Air Act upon amendment in 1977. Still there was no express reference to ecological considerations. Congressional focus remained on anthropogenic changes in the stratosphere that could "cause or contribute to endangerment of the public health or welfare". The need to infer ecological health only in the name of human welfare—or health—lingered.

Future direction is, however, at least suggested in the congressional finding to the effect that increased solar ultraviolet radiation as a result of "holes" to stratospheric ozone could increase human disease rates, threaten food crops, and most significantly to ecological concerns, "otherwise damage the natural environment".

Improved attention to ecosystems is more promising in the charge to the National Science Foundation calling for research intended to "increase scientific knowledge of the effects of changes in the [stratospheric] ozone layer upon living organisms and *ecosystems*".

Ecosystems, Air, and the Amendments of 1990

When the Clean Air Act underwent major amendment in 1990, a substantial increase in the ecological sophistication of Congress was suggested. Definitions began to accommodate ecological features directly as well as indirectly through the public welfare route, research projects commenced to show signs of incorporating ecological realities, matters of more ecological than merely human welfare interests emerged, and regulatory programs took on an ecological component beyond any that happened to be associated with ambient air quality and designated pollutants.

The most obvious general incorporation of ecological considerations can be found in the Programme addressing hazardous air pollutants. New subsection adds a mandate for regulating as hazardous an emission posing any significant and widespread adverse effect, which may reasonably be anticipated, to wildlife, aquatic life, or other natural resources, including adverse impacts on populations of endangered or threatened species or significant degradation of environmental quality over broad areas.

Additionally, "welfare" throughout the statute is expanded to include effects on soils, water, vegetation, animals, wildlife, weather, and climate. The 1990 amendments do provide expressly and directly for ecosystem research, specifically "to improve understanding of the short-term and longterm causes, effects, and trends of ecosystem damage from air pollutants".

Considerable sensitivity to complexity in ecosystems is discernible in congressional direction to address causes and effects associated with chronic and episodic exposures and calling for a determination of reversibility. The latter also shows some acceptance of resiliency in ecosystems—the characteristic tendency to restore homeostasis upon perturbation. Research efforts are also directed to promote improved understanding of multiple environmental stresses associated with air pollution and, showing still greater sophistication, evaluation of effects on water quality attributable to acid deposition and other atmospheric pollutants.

There is reference to wetlands, estuaries, and groundwater, all of which are or affect ecosystems that may be sensitive to airborne chemical substances. Terrestrial ecosystems are not neglected. There is reference to soils, forests, and biological diversity. Certain ecological matters that had not been amenable to action under the Clean Air Act are expressly brought into the Programme.

The ecological significance of stratospheric ozone and the corresponding basis for protective measures are assumed in section 602 in which ozonedepleting anthropogenic chemicals are listed for the purpose of monitoring and reporting and actual elimination from production and consumption in the marketplace. There is even anticipation of a potential call for an accelerated schedule on behalf of the environment as well as of human health.

Acid deposition is the subject of continuing research, with particular reference to continuation of programs established in the Acid Precipitation Act. Research, focusing on water quality, forests, and soil, accommodates time scales in keeping with those at which ecosystems are functioning. More general deposition of air pollutants in the Great Lakes, Chesapeake

Bay, Lake Champlain, and coastal waters is addressed as part of the vastly expanded National Emission Standards for Hazardous Air Pollutants (NESHAPS) Programme. Among other things, the relevant subsection calls for the sampling of pollutants in the "biota" and in fish and wildlife.

Biota goes undefined. In the cases of Chesapeake Bay and Lake Champlain, an expansion of interest to include watersheds suggests recognition of how one ecosystem can merge into another and of the interconnections between aquatic and terrestrial ecosystems.

At a more general level of regulation, the statutory frameworks for criteria pollutants as well as for hazardous pollutants now take into account ecological amenities above and beyond any that may be brought in through the back door of human welfare.

In revising the congressionally established list of hazardous pollutants, the EPA may by rule address environmental effects resulting from ambient concentrations of those pollutants, from deposition and bioaccumulation and other processes-a refreshing acknowledgment that all the possibilities cannot be predicted from the splendid isolation of congressional offices.

Ecosystems and the Clean Water Act

The Clean Water Act, as amended through 1990, followed the Clean Air Act in more than the chronological sense. The concept of ambient air finds its counterpart in water quality, while discharges to surface waters correspond to emissions into the air. In the case of water, however, ecological components were from the outset a major component of the statutory framework.

Since the contemporary origins of federal water pollution control, ecological systems have increasingly been integrated into emerging policy. They are found in definitions, research programs, and regulatory regimes.

Taken altogether, the relevant provisions of this statute reveal attention to the flow of energy, material cycling, strategy and homeostasis, population and community interactions, and even more subtle components of the generic ecosystem as manifested in aquatic habitats. But even as this improved attention to ecology takes hold, a lingering anthropocentrism remains.

An ecological foundation is present from the opening of the Clean Water Act, with the statement of one major objective: the restoration and maintenance of the nation's waters in terms of physical, chemical, and biological parameters. A related interim goal is water quality suitable for the protection and propagation of fish, shellfish, and wildlife.

Statutory definitions suggest more than a passing interest in ecosystem components. For effluent standards, a toxic pollutant is defined through the capacity upon exposure, ingestion, inhalation or assimilation into any organism, either directly from the environment or indirectly by ingestion through food chains,... [to] cause death, disease, behavioral abnormalities, cancer, genetic mutations, physiological malfunctions [including reproductive] or physical deformations, in such organisms or their offspring.

This one definition reflects increased ecological awareness. Trophic levels of simple food chains or more complex food nets are recognized, as are mortality in one or more individual organisms and in whole assemblages. The more subtle reality of the importance of routes of entry breaching the integrity separating internal self from the external environment is accepted.

Further, the requirement that environmental components be in contact in order to interact is also accepted. Finally, perturbation at the organismal level is acknowledged with reference to an array of biological endpoints ranging through behavioral abnormalities, physical deformation, altered biochemistry, chronic toxicity, congenital defects, and inheritable damage.

At another level altogether, the definition of biological monitoring first of all recognizes a basic intent of the Clean Water Act to address ecosystem needs and then factors implying compromise of those needs, such as accumulation of chemical substances in tissues.

This definition also recognizes certain ecological realities in calling for sampling at appropriate levels in the food chain with reference to the "volume and the physical, chemical and biological characteristics of the effluent" and to "appropriate frequencies and locations," in acceptance of all four significant dimensions.

Sensitivity to ecological features is found with reference to planning for reservoirs: "The need for and the value of storage for regulation of streamflow (other than for water quality) including but not limited to salt water intrusions and fish and wildlife, shall be determined" by responsible agencies. Planning is encouraged to include "basins," that is, "rivers and their tributaries, streams, coastal waters, sounds, estuaries, bays, lakes, and portions thereof as well as the lands drained thereby". Actual natural systems are addressed as whole entities.

Among them are the Great Lakes System, which by definition includes the entire drainage basin of the Great Lakes themselves, a clear recognition of the necessity to expand the boundaries of an ecosystem from the microcosm of immediate regulatory interest through ever-expanding boundaries to include all the natural sources of contribution.

The statute addresses estuaries with particular attention, first in defining an "estuarine zone" as "an environmental system" including the estuary itself and "transitional areas which are consistently influenced or affected by water from an estuary."

An estuary itself is defined as that portion of a river or stream "having unimpaired natural connection with open sea and within which the sea water is measurably diluted with fresh water derived from land drainage".

The corresponding research Programme speaks in terms of balance and of indigenous organisms. Trends are to be assessed. Both shortterm and long-term conditions are taken into account. Among the research efforts are those directed to ecosystems, water quality, and designated or potential uses of estuarine systems. The last factor is the most obviously directed to human interests.

Research efforts are also directed toward identification and classification of the eutrophic condition of all publicly owned lakes. Although eutrophication is the natural process through which every lake proceeds, accelerated eutrophication is a nuisance situation attributed to the presence of excessive nutrients associated with human activities.

Congress established River Study centers to conduct the interdisciplinary studies essential to comprehending ecological features in relationship to human needs including usage and economics. Here there is reference to the ecological phenomenon of bioaccumulation, albeit in the human contexts of "reducing the value of aquatic commercial and sport industries" and restoring and enhancing such "valuable resources".

In sum, members of Congress have demonstrated some comprehension of how chemicals move through and affect ecosystems. Organic chemicals, for example, may enter an ecosystem and serve as "exotic" nutrients or interfere with the intricate metabolism and catabolism of the biotic components of the system. Heavy metals are micronutrients, but slight excesses beyond what organisms require in their physiological processes will be toxic in terms of subchronic, chronic, or acute alterations in those processes.

Congress focuses, however, on a species of immediate human interest, which may be seriously affected only after a long interval of more subtle harms with increasingly apparent biological damage, perturbation.

As should be expected, actions on behalf of ecosystems are most apparent in establishing water quality standards. Standards and associated implementation plans are directed to include enhanced water quality—first and foremost an ecological matter, even were the interests of a sole species, *Homo sapiens sapiens,* taken into consideration.

But section 303 goes at least slightly beyond immediate human interests, with references to propagation of fish and wildlife and to thermal loads, the latter in the contexts of such factors as normal water temperatures, flow rates, and seasonal variations.

Here, the Congress is acknowledging the crucial fourth dimension in ecology. Recognizing that each ecosystem is somehow unique, Congress has addressed specific water quality needs of specific bodies of water.

Effluent standards are not so directly connected with ecosystems, although they are in part directed to ensuring that designated water quality standards are met. There is an exception in the so-called guidelines on which permits to discharge into ocean waters must be based. These guidelines are intended to determine degradation of waters and make specific reference to communities of organisms, notably including plankton.

Additional references are to changes in marine ecosystem productivity and stability and alterations in communities' composition. In doing so, Congress implies a recognition of the varying significance of absolute amounts, concentrations, and rate of entry into a system.

Ecosystems return to the foreground in the liability provisions of section. Liabilities associated with spilled oil or hazardous substances include monies to be reimbursed to a natural resources trustee for replacement or restoration of such resources.

Section speaks in terms of harmful quantities and mitigation with reference to environmental amenities with express inclusion of organisms and their habitats within the aegis of "public health or welfare", leading the way to recognition of *Homo sapiens sapiens* as an ecological entity in need of ecosystems that are healthful to its populations.

In some senses, it is only a short next step to recognizing a human requirement for ecosystems that are healthy in terms of their own immediate needs.

ADDITIONAL POLLUTION CONTROL PROGRAMMS ADDRESSING ECOSYSTEMS

The Toxic Substances Control Act, the Resource Conservation and Recovery Act, also known as the Solid Waste Disposal Act, and the Comprehensive Environmental Response,

Compensation, and Liability Act, also known as "Superfund," all followed the Clean Air and Water acts in the course of the 1970s, the first decade of the environment, with RCRA's hazardous waste regulations and CERCLA both taking effect beginning in 1980. Of the three, the TSCA pays the most direct attention to ecological amenities.

Assuming that first in reference is first in importance, Congress ranked the environment second only to human beings in its findings of exposure to and unreasonable risks attributed to chemicals in commerce. This paired set of priorities is maintained throughout the statute. Environment is defined to include "water, air, and land and the interrelationship which exists among and between water, air, and land and all living things".

The kinds of toxic properties of interest pursuant to the TSCA in terms of environmental amenities expressly include "carcinogenesis, mutagenesis, teratogenesis, behavioral disorders, cumulative or synergistic effects." Additional characteristics of concern are "persistence, acute toxicity, subacute toxicity [and] chronic toxicity". These terms are not exclusively bound to human health but expressly include other biological systems.

Similarly, the RCRA calls for "environmentally safe disposal of nonrecoverable residues" (solid wastes) and associates environment closely with human health. The hazardous component of solid waste is defined as such in part by reason of its detrimental effect on the environment (second only to human health).

Ecosystems enter CERCLA by way of the natural resources route. Damages, as defined by this act, are specifically those "for injury or loss of natural resources", which are defined to include land, fish, wildlife, biota, air, water, and groundwater.

Environment is that which is so defined pursuant to a variety of other national programs but essentially including all environmental media. Hazardous substances include all those that can adversely affect environmental amenities, as defined in CERCLA's predecessor pollution control statutes.

A pollutant or contaminant is defined in terms of environmental perturbations as well as of human health concerns. The language of the definition follows that found in section of the Clean Water Act.

CERCLA's Programme of strict, joint, and several liability includes liability for damages to natural resources. This provision, however, sheds no further light on congressional understanding of ecosystems as affected by oil and hazardous materials. Section calls for two sets of procedures for determining the extent of harm.

One is to establish minimal "simplified assessments requiring minimal field observation," while the other is to establish a more elaborate protocol "to determine the type and extent of short-term and long-term injury, destruction or loss." Somehow, responsible bureaucrats must consider such factors as "replacement value, use value, and the ability of the ecosystem or resource to recover."

NATURAL RESOURCES MANAGEMENT

Because statutes addressing natural resources directly are obviously more closely linked ecological principles than are pollution control statutes, the codified foundations of natural resources law can be expected to be more accommodating of those principles. Here, that expectation will be pursued in congressional findings, missions, and definitions in selected statutes..

The declaration of policy of the Multiple-Use Act calls for administration of national forests for purposes of "outdoor recreation, range, timber, watershed, and wildlife and fish". Only the final three concerns bear any obvious relationship to ecological amenities as opposed to resources of value for human use, although the call for establishment and maintenance of wilderness areas is at least promising. Unfortunately for ecological amenities, even reading between the lines of the definition of multiple use reveals little ecological sensitivity.

Among the declarations of congressional policy in the Land Management Act is that calling for management in a manner that will protect the quality of scientific, scenic, historical, ecological, environmental, air and atmospheric, water resources, and archeological value; that, where appropriate, will preserve and protect certain public lands in their natural condition; that will provide food and habitat for fish and wildlife and domestic animals; and that will provide for outdoor recreation and human occupancy and use.

Although this is a mixed mission well down the list of missions, it does allow for inferences of ecological amenities with resort to reading between the lines.

The import of statutory definitions is similarly mixed, with special reference to the definition here of multiple use to include the implicit, and somewhat cryptic, direction to manage public lands so that they are utilized in the combination that will best meet the present and future needs of the American people; making the most judicious use of the land for some or all of these resources or related services over areas large enough to provide sufficient latitude for periodic adjustments in use to conform to changing needs and conditions.

FEDERAL PROGRAMMS BASED IN THE ECOSYSTEM CONCEPT

At least two national programs are expressly directed to resources that can be described as ecosystems. One of these programs is statutory, the Coastal Zone Management Act. The other must be extracted through regulatory definitions of wetlands. The point of departure for the former is found in coastal ecosystems, whereas a major focus of the latter is the concept of habitat. Because the focus here is on congressional recognition of ecological concepts, only the statutory Programme will be addressed.

The wetlands Programme remains extraordinarily controversial and unsettled at the point of regulatory programs and their implementation.

Explorations for ecological sophistication in the CZMA will be limited to congressional missions and definitions. Congress found that the coastal zone is rich in ecological as well as other features and recognized that the competing demands of coastal lands and waters had led to "loss of living marine resources, wildlife, nutrient-rich areas, [and] permanent and adverse changes to ecological systems". Congress also recognized the fragility of associated habitats and, perhaps more important in terms of public policy, that ecological values are among those that are essential to the well-being of all citizens.

The resultant policy includes protection of ecologically meaningful natural resources such as "wetlands, floodplains, estuaries, beaches, dunes, barrier islands, coral reefs, and fish and wildlife and their habitat".

Just how aware of ecological principles these codified materials are will become increasingly apparent as those principles are explored throughout the remainder of the text. Upon return to matters of policy, it should become easier to read through the legal language to find and assess ecological foundations. Every return should be guided by the question: Is there such a biological entity as described by the law?

Chapter 4

Structure of Ecosystem

It is an inherent paradox of science that the holistic concept of ecosystem can actually be expanded through a reductionist approach supplied by physics and chemistry. Although the ecosystem is more than the sum of its constituents, details of interactions between and within the abiota and biota provide direction and insights into the structure and function of the whole.

Planet Earth was created and itself evolved out of the energy and material of the Big Bang. Life arose after ages of geological evolution leading to conditions amenable to the coalescence and persistence of certain complexes of organic molecules. In a continuing process, the biota of an ecosystem is initially determined by these abiotic features as they are uniquely combined in locations throughout the globe.

Because it is energy that drives the whole system, the energy environment will be explored first. Then material cycling between abiotic reservoirs and pools and the biota and within organisms will then be blended intc the ecological mixture.

THE ENERGY ENVIRONMENT

Energy enters the ecosystem as light and exits as heat. Energy takes numerous forms, few as obvious than the dropping of a rock off a precipice. In the biosphere, light, a portion of the electromagnetic spectrum, is kinetic energy, as are the wave particles of that entire spectrum. Mechanical energy is potential in a relaxed muscle and kinetic in the muscle contracting to move a limb or blood through the circulatory system. Chemical energy is the means whereby organisms absorb and assimilate the energy in light for work in the maintenance, growth, functioning, and reproduction of living entities.

THE LAWS OF THERMODYNAMICS

In order to understand the significant detail of energy flowing through an ecosystem, it is necessary to have a modest grasp of certain realities of physics. Outstanding among these are the First and Second Laws of Thermodynamics.

The First Law holds that energy and matter can neither be created nor destroyed. In essence, that reality has held true at least from the instant following the Big Bang. All the energy on earth can be traced to the thermonuclear reactions in the sun, present and past.

Energy flow is mediated through conversions among types of energy with the production and rearrangements of material in cells. Similarly, preexisting material is metabolized and catabolized within organisms, with changes in chemical structure without actual material creation. In an ecosystem—or any biological entity—preexisting energy and material are fixed in a form that the biota can assimilate; energy is converted into different forms, as is matter; and the two are exchanged: Energy is converted to matter, and matter to energy.

Throughout all these processes, nothing is created out of nothing. There are specific energy and matter precursors to everything. The energy comes from the sun, and the material comes from the atmosphere, hydrosphere, and lithosphere making up the planetary biosphere. Along the way, however, energy is continually dissipated, in accord with the Second Law of Thermodynamics.

There are a number of relevant ways in which this concept can be stated for ecologists. The crucial point is that energy always tends to dissipate from a higher to a lower or more dispersed concentration. A cup of hot coffee left to its own devices never becomes hotter; it always cools to room—ambient—temperature. A sugar cube in a beaker of water dissolves, and its molecules disperse throughout the liquid. Those molecules will not reunite to re-form the sugar cube. This persistent loss of energy (organization or order) is measured as entropy (randomization).

In any undisturbed system, particles (including individual chemical elements or molecules) will become as disordered as possible. Their distribution is random, the antithesis of order. Life in thermodynamic terms is a kind of negative entropy.

Additional versions of the Second Law are to the effect that no energy transfer is 100% efficient; there is always some loss, usually as heat. Thus, energy is increasingly dispersed, decreasingly available for use in the system. Finally, no process involving an energy transfer will occur spontaneously unless there is a degradation of energy content from start to finish. Chemically, the products will contain less energy than the reactants.

Electromagnetic Radiation: The Energy Spectrum

Electromagnetic radiation, particularly that portion of the energy spectrum identified with visible light, is the foundation of the energy environment in which organisms function. Radiation is defined as the emission or propagation of energy in waves through space or material. Each wave in the spectrum is defined in terms of amplitude, frequency, and wavelength. Amplitude is the height of each wave; frequency, the number of waves generated per unit time; and wavelength, the distance between peaks.

Although all the parameters are interrelated, for purposes of ecological significance, the spectrum is described in terms of wavelength.

The visible portion of the electromagnetic spectrum ranges from the short wavelengths in the ultraviolet through the long wavelengths of the infrared. The longer wavelengths in the ultraviolet range remain visible to humans, but the shorter wavelengths become invisible and, indeed, are disruptive to cellular components, notably the genetic material. Ultraviolet radiation is mutagenic. From the ultraviolet, the spectrum shortens to X rays, which also are mutagenic. At the longer extreme of the spectrum, some infrared remains visible to humans but becomes invisible as wavelength approaches microwaves and radio waves.

The visible range bears a significance beyond human vision. It is within this range that biological molecules can be generated and persist for durations suitable for metabolic processes. Shorter wavelengths are too energetic and disrupt biochemical molecules. Longer ones lack sufficient energy to enter into the processes of life.

All the sun's electromagnetic energy entering the biosphere is reflected, absorbed, or converted to heat. All these fates contribute something to the planet's capacity to support life. Much light is reflected from the various surfaces it strikes. Certain wavelengths of light are absorbed by the biota and assimilated to conduct the work of the ecosystem.

Others are absorbed by surrounding water or ground, contributing to temperatures in which living processes can occur. The remainder are converted to heat, which is dissipated to the surrounding environment, through the atmosphere beyond the planet, through the solar system, and ultimately, into outer space—continuing on the cosmic pathway to the entropic destiny.

In quantitative terms, sunlight reaches the biosphere with an energy content of 2 calories for each square centimeter per minute. As the light penetrates the atmosphere, much of the energy is lost. Upon reaching the surface of the earth, the energy content is reduced to 1.34 calories for each square centimeter per minute.

That maximum quantity is said to exist at noon on a clear summer day. Additional attenuation is virtually certain through the presence of clouds, because of passage through water, or because of absorption by vegetation. Moreover, the sun's radiation occurs for only part of the diurnal period.

The difference between downward and upward streams of radiation is expressed as net radiation. Less radiation is released than is received because of general dissipation and also because of the evaporation of water and the generation of thermal winds. The net transfer of heat serves to maintain the temperature range in the atmosphere that is conducive to life. Some of that light performs work or is converted to forms that themselves perform work. Prominent examples are the winds and the flow of the hydrogeological cycle. A small portion of the light is absorbed by chemical receivers in the cells of certain organisms, and this energy is converted into biomass.

ENTROPY AND LIFE

None of the work accomplished by the light received from the sun, according to the Second Law of Thermodynamics, can fully utilize the contained energy. This unused energy is converted to heat, the vibration of molecules measured as temperature. The faster the vibration, the hotter the temperature.

The heat of the entering energy and that being released in the course of chemical reactions (a form of work) serves members of the biota as part of the environment affecting biochemical reactions. The ecosystem is, in short, driven by the energy of the sun, but it is a highly inefficient process. Together, these two realities of physics in combination answer the biological questions of why elephants are the largest terrestrial animal and why this organism is hairless and endowed with large ears.

The elephant represents the maximum limit on terrestrial size because the heat of biochemical reactions in a larger animal could not be dissipated through the mass of elephant matter quickly enough to avoid fatal overheating at the cellular, and ultimately the organismal, level. Elephants are hairless to allow the dispersion of internal heat through the skin without the insulating effect of a fur coat. The large ears are a medium through which additional heat can be released, especially when they are flapped to enhance that release.

In the case of particularly large plants, metabolic processes are occurring only in a small portion of the whole. This limitation is obvious in the lack of leaves among the lower branches of huge trees. It would be wasteful of energy for trees to grow and maintain leaves in shaded areas where photosynthesis would be inhibited. Stated slightly more precisely, growth and maintenance would require more energy from the plant than the shaded leaves would be able to provide it.

Other adaptations for controlling energy flow in extreme climates are quite striking. In deserts, plants are relatively compact and low growing, with thickened stems, whereas animals tend to be small but lanky with prominent ears. In cold regions (alpine or subpolar), plants and animals are also small and far more compact than their desert counterparts. Animals have stubby ears, close to the body. In both regions, the relatively large ratio of surface area to mass limits the entry or loss of energy as heat. In addition, large ears radiate heat to cool the body; small ones help retain body heat.

ENERGY FLOW THROUGH THE ECOSYSTEM

The relationships among the components of an ecosystem are governed by the same natural laws that apply to nonliving systems. A major difference is in the degree of organization required for the phenomena making up living systems. From the cell to the biosphere, each succeeding level reflects greater organization than that below. Life overall is a momentary defeat of entropy. As such, life requires extraordinary amounts of energy, a condition that has been identified in terms such as negative entropy or an energy pump.

PHOTOSYNTHESIS AND RESPIRATION

Photosynthesis. For ecological purposes, the effective flow of energy commences with its biological fixation. If the energy of the sun simply struck members of the biota with nothing more than absorption or reflection, there would be no life. These are not effective events in terms of biological processing. Instead, a specific kind of assimilation takes place. Green plants absorb in most of the visible spectrum (excluding the greens) and in the far-infrared, the longest wavelengths of light.

Chlorophyll, the molecule of photosynthesis, absorbs more particularly in the blue and red. The energy contained in these wavelengths is biologically fixed through photosynthesis, actually a series of biochemical reactions. Those collective reactions are summarized in the formula describing the reactants and ultimate products:

$$6CO_2 + 6H_2O \xrightarrow{light} C_6H_{12}O_6 + 6O_2$$

What the formula stands for is the combination of six molecules of the gas carbon dioxide plus six molecules of water, in the presence of light energy, for conversion into one molecule of the sugar glucose and six molecules of gaseous oxygen.

The first of the steps making up photosynthesis is the one requiring light energy, and it occurs in the presence of chlorophyll, found in cellular organelles of uniquely characteristic structure known as chloroplasts. The remaining reactions are known as dark reactions because light is no longer required.

Respiration: The apparent chemical reversal of photosynthesis is respiration:

$$(CH_2O)_n + nO_2 \rightarrow nCO_2 + nH_2O + \text{energy}$$

If the carbohydrate is a six-carbon sugar, such as the glucose formed in photosynthesis, "n" becomes "6," and the generalized formula becomes specific:

$$6CH_2O + 6O_2 \rightarrow 6CO_2 + 6H_2O + \text{energy}$$

This formula translates as one molecule of sugar reacting with six molecules of oxygen to release six molecules each of water and the gas carbon dioxide, with the release of the energy stored in the molecular bonds holding the elements of the sugar together.

Together, photosynthesis and respiration relate the processes whereby green plants capture and store energy and all organisms, including green plants, release that energy for use in their cells. It is worth noting at this stage that photosynthesis has a biochemical counterpart that requires no oxygen. That process is mediated by photosynthetic bacteria that use sulfur instead of oxygen: SS

$$12H_2S + 6CO_2 \rightarrow C_6H_{12}O_6 + 12S + 6H_2O$$

This formula is useful in demonstrating how the photosynthetic reaction occurs. It is the oxygen of carbon dioxide that is incorporated into the sugar. The oxygen of water, like the sulfur of hydrogen sulfide, is that which is released into the environment.

TROPHIC LEVELS: FOOD CHAINS AND FOOD WEBS

Clearly, the physics and chemistry of life are not readily isolated from each other. It is difficult to consider one without resort to aspects of the other. The same is true of life. While living organisms are much more than particularly well-organized sets of chemical reactions and physical processes, it is difficult to isolate any one of the three pieces from either of the other two.

Along the way from the entry of light into an ecosystem and the release of entropic heat, the energy is flowing through the biota while matter is cycling among all the components. There is a striking structure to this functioning whole, which will be explored further when the distribution of materials is superimposed on the flow of energy.

Each trophic level is made up of populations of organisms that have precisely defined places in the whole. Taken altogether, with identification of the taxons actually participating, these constituents of energy flow describe not merely some generic ecosystem but a specific one at a specific stage in its life history.

The Grazing Cycle

The first trophic level in the grazing cycle is that of the primary producers, the green plants. These organisms are the ones that fix light energy, carbon dioxide, and other inorganic substances into biologically useful forms. They are autotrophic; that is, they can convert inorganic material into organic. In fact, they cannot utilize the organics produced by other organisms.

The second trophic level is that of the primary consumers. These organisms, "grazers," are the ones that feed directly on the producers. In biological terms, primary consumers are heterotrophic. Unable to produce needed organic molecules from inorganic, they convert those generated by the producers to the biochemical molecules they need. Primary consumers are known as herbivores and are among the phagotrophs. They ingest relatively large pieces of whatever plants serve their metabolic needs. Many of these consumers are obligate herbivores; their physiology cannot be served by anything but the plants on which they feed.

The third step in this food chain of trophic levels is the secondary consumers, the first level of carnivore. These organisms feed on primary consumers, the herbivores. They, too, are heterotrophic, but their metabolic processes require organic materials as they have been rearranged by their prey organisms. The carnivores are also phagotrophs, and many are obligate carnivores.

The next and succeeding trophic levels are carnivores that feed on carnivores, but there is a limit to how many levels can be supported by an ecosystem. That limit is related to the Second Law of Thermodynamics-hence, the trophic pyramid, with biomass decreasing from the producer base to the tertiary or quaternary consumer at the top.

Organisms in the food chain, true to the First Law of Thermodynamics, are assimilating energy and using it to assimilate matter and to rearrange its organization to meet their respective physiological needs. All the processes are subject to the Second Law of Thermodynamics. Increasingly large proportions of energy are lost to entropy as it proceeds through the biomass from one trophic level to the next. Ultimately, it will require more energy than the food organism can provide, just to get it and assimilate it.

The proportion of biomass bound up in each level decreases. In fact, some 80% to 90% of the potential energy is lost to entropy at each step (energy transfer). Either numbers are reduced or volume is, and the number of trophic levels is ordinarily limited to no more than five. An obvious corollary is that the shorter the chain, the greater the available energy at the top. Hence, big, fierce animals are rare in any ecosystem. And trophic steps are usually limited to four or five at the most. Actually, there are two major complications here. First, some organisms are omnivores and others are facultative herbivores or carnivores. Omnivores have the biochemical apparatus to allow them to feed on both plants and animals, with a triggering of the requisite metabolic processes for assimilating the various organics being ingested.

Facultative organisms can vary their diets through alternative metabolic pathways, turned on or off depending on the diet of the moment. The second major complication is that most real ecosystems are not limited to mere food chains. Instead, there are food nets, with organisms readily feeding at more than one trophic level and numerous predatorprey interaction.

The Detritus Cycle

Because the grazing cycle is so immediately obvious, it is easy to neglect the detritus cycle mediated by detrivores. This is the cycle by which the elaborate biochemical molecules generated through the grazing cycle are restored to the abiota and become available to reenter the grazing cycle.

The detritus cycle begins with consumers of carrion and of dung. Much of this cycle is mediated by series of insects and soil microorganisms. The latter, in particular, are quite invisible to the casual observer, but without them, the full material cycle would come to a halt.

While all of the organisms in the detritus cycle are heterotrophs, some are phagotrophs and others are saprotrophs. As organic materials proceed through this cycle, they are chemically converted in distinct stages to the inorganics from which the grazing cycle metabolized them.

For each link in a food chain, energy is partitioned in a manner suggesting the other links in the chain. Ingestion represents the full amount of energy available to the organism. This is potential energy bound up in the material of the ingested biomass.

Much of the energy is expended in the chemical processes of digestion and assimilation. In these and other cellular processes, respiration can be described as the energy used to perform the work or lost as heat and no longer available to the community. That material that has been assimilated is converted to growth and reproduction—additional biomass in the form of the ingesting organism and its offspring. Additional energy is returned to the ecosystem throughout the life of the ingesting organism through the processes of egestion, excretion, and finally, death. Egested materials are essentially unaltered chemically and play no positive role in metabolism. Significantly, energy is required to process this material, but it contributes none to the organism.

Egestion is an energy economy in that the material unavailable to the organism is eliminated prior to entering the metabolic processes of digestion and elimination. Owls egest the inedible portions of their prey, such as fur, skin, and bones. The resultant pellets contain these materials virtually intact.

Excreted materials are the chemical wastes produced in the course of metabolic processes. Were these materials retained, they would eventually poison the organism. With death, all metabolism ceases, including defenses against detrivores. Every release of material carries energy to the detrivores.

THE MATTER OF PRODUCTIVITY

For a variety of reasons, including determinations of the condition of an ecosystem, a measurement of productivity may be in order. Generally, productivity can be defined as the amount of energy received from the sun that is retained in the biomass of the ecosystem over some specified interval of time measured in days, months, or years.

There are four ways to account for productivity. Gross primary productivity measures the total rate of photosynthesis, the full assimilation of energy by primary producers. Net primary productivity is the rate of energy storage as plant biomass in excess of respiratory utilization. This is "apparent" photosynthesis or "net" assimilation.

Net community productivity is the rate of storage of organics not used by heterotrophs—plant biomass left untouched. It is calculated by subtracting heterotrophic consumption from net primary production. The fourth accounting, secondary productivities, includes rates of energy storage at the several consumer levels. Production has been replaced with conversions. High rates of production are expected under favorable physical conditions, internal to the ecosystem and in the form of energy subsidies from outside the ecosystem. Wind and rain are energy subsidies entering rain forests. Tidal energy serves estuaries. Such subsidies are, however, not limited to naturally occurring ecosystems. They are vitally important factors in cultivated systems as well.

With agricultural ecosystems (often monoculture crops of little genetic diversity), high rates of production are related to energy subsidies from outside the system. Cultivation has always called for work beyond the immediate assimilation of solar energy.

First, human Labour is required, historically supplemented with that of animals. In recent times, both have been replaced to varying degrees by the fossil fuels needed to run laborsaving agricultural equipment.

The important consideration to be taken from measurements of productivity is that mere weight of biomass and counts of organisms are insufficient measures of ecosystem health. Here carefully selective quantitative measurements can provide a measure of quality. Any such effort, however, must take into account the presence of energy drains, on one hand, and subsidies, on the other. Energy drains are stresses diverting energy from the system. A familiar example is the removal of cereals, fruits, and vegetables from agricultural crops. The nature and magnitude of such important ancillary phenomena must be included in the description of energy flow and retention.

QUANTIFYING PRODUCTIVITY

The ideal quantification of productivity would directly measure the flow of energy through the system. To do so, unfortunately, is difficult at best. Thus, ecologists turn to indirect methods of measurement. Significantly, no two methods measure exactly the same aspect of metabolism. Using more than one can serve as confirmation of both. Seven methods of quantifying productivity are available. The most obvious is:

- Harvesting of the crop. Also obvious are measurements
- Oxygen
- Carbon dioxide changes and of the
- Disappearance of raw materials. Additional methods refer
- pH, the use
- Radiotracers,
- Chlorophyll.

The harvest method measures the weight of growth from seed (set at zero) for a determination of caloric content. It is used in cultivated crops and wherever annual plants (in contrast to biennials and perennials) predominate. The method is also used in early stages of revegetation of damaged ecosystems.

The limitations of the method are not surprising. It is inappropriate where herbivores are important and actually present, where steady state has been reached, and where food is removed as it is produced. (Steady state is the condition of relatively long-lived homeostasis in a mature ecosystem.) When herbivores, including humans, are consuming or removing plant material, there will be direct and indirect effects on the rate and final extent of productivity. Once steady state is reached, productivity will diminish and cease.

Oxygen measurements compare the presence of this gas in closed vessels representing gross primary production and net community production. The amount of oxygen depletion is proportional to productivity.

This system is limited to aquatic systems and cannot measure metabolism in depths beyond the photosynthetic zone. Because the community being analyzed must be contained in small vessels to allow the measurements, an unknown experimental factor is the effect of such containment. There is no guarantee that the experimental system is fully comparable to its natural counterpart.

Carbon dioxide methods measure either the uptake or the release of that gas, thus measuring net photosynthesis or respiration. The method is limited to enclosed terrestrial systems. Another limitation of the method is that the enclosure chamber can act like a greenhouse and allow the buildup of heat, which can alter the rate of photosynthesis. Moreover, the sheer size and complexity of most terrestrial ecosystems render enclosure difficult.

The pH method measures acidity as a function of dissolved carbon dioxide content and is thus limited to aquatic systems. Concentrations of this gas will decrease with photosynthesis and increase with respiration, with corresponding alterations in pH. However, this method is of limited application because water bodies have differing buffering capacities, and there is no linear relationship between pH and carbon dioxide content. A calibration curve must be established for each analysis to reflect the actual relationship between the two parameters.

If the biota of interest is not in a steady-state equilibrium, the disappearance of nutrients, for example, minerals, will provide a measure of productivity. This measurement, however, will be influenced by a whole host of additional abiotic factors, which may or may not be taken into account or even recognized.

Radiotracers, radioactive isotopes of biochemically active elements, will provide more information regarding distribution of those elements in and among cells, tissues, organisms, and the ecosystem at large than rates. Finally, the chlorophyll method assumes with some degree of logic that the more photosynthesis is occurring, the more chlorophyll will be present.

IMPLICATIONS FOR ENVIRONMENTAL PROFESSIONALS

Productivity, naturally occurring or in agricultural or other ecosystems subject to human manipulation, if properly measured, can be informative to the environmental professional. Studies in research and applied ecology reveal certain expectations in the energy relationships in specific ecosystems. Altered relationships suggest the presence of unrecognized energy subsidies or drains.

It is the environmental professional who may be called on to identify the external condition in order to determine whether the effect is amenable to correction through the marketplace or public intervention.

MATERIAL CYCLING AND ECOSYSTEM STRUCTURE

Driven by energy, the function of an ecosystem is reflected in the structure assumed by its community of organisms in their niches that fill the trophic levels. These trophic levels also indicate the distribution of material. The resultant relationships reveal a striking degree of organization that lends itself to simplification for purposes of analysis conducted by ecologists or environmental professionals.

With the flow of energy, abiotic dissolved nutrients are biologically fixed as they are assimilated into plant biomass. This biological material is eaten by herbivores that indulge in their individual versions of assimilation and biochemical conversion of the plant biomass.

Herbivores are eaten by carnivores, which may be eaten by still other carnivores, each with its own unique pattern of assimilation and conversion.

This grazing cycle restores certain nutrients through excretion and returns complex organics to the abiota through the formation of detritus. The detritus cycle consists of phagotrophs and saprotrophs (bacteria and fungi) that decompose the complex organics, taking these molecules stepwise back to their inorganic components, notably water and carbon dioxide from which the nitrogenous and phosphoric compounds and trace metals are released. The ultimate products of the detritus cycle are the dissolved inorganic nutrients to be assimilated by plants.

Additional connections exist between these two cycles and the abiotic environment. Detritus feeders and those that feed on bacteria and fungi are themselves consumed by certain carnivores. Symbiotic microorganisms may work in combination with plant hosts to assimilate certain nutrients directly.

Autolysis, the internal enzymatic breakdown of cells, also returns dissolved nutrients to the environment, as does microbial hydrolysis. It is within this structure that the chemistry of life is taking place.

In moving from energy flow through the ecosystem to material cycling within it, the opening question might well be the fate of the glucose generated as a result of photosynthesis. In fact, the glucose is biochemically related to the Krebs citric acid cycle, the chemical centre of respiration, described below. The mildly expanded and revised shorthand for the complex of reactions involved is in the two formulas:

$CO_2 + H_2O$ + minerals (inorganic nutrients) $\xrightarrow{\text{light}}$ Plant Production

O_2

Transpired water

Sugar → biomass (growth and reproduction)

Biosynthesis

Maintenance

CO_2

It is highly informative now and for future reference to note that in the first formula the oxygen and water are in some senses by-products, whereas in the second, carbon dioxide is the by-product. *By-product* is legitimately translated "waste."

In ecological systems, one organism's waste is another's source of nutrients. Nevertheless, either too much waste or too much of a nutrient can be detrimental to the system and its living components. Similarly, oxygen or carbon dioxide, as representative of all essential nutrients, in the *wrong* place, that is, in the immediate environment of certain organisms, can be toxic. How a substance is going to affect an ecosystem depends entirely on its chemical nature, absolute amount, concentration, and location. The biochemical molecules needed for all the processes of life are generated in the cells from the products of photosynthesis and respiration. The molecules of life start with simple sugars and the organic acids of the Krebs cycle, but these basic patterns are elaborated into amino acids and proteins, and an array of lipids including fats, oils, steroids, and waxes.

In fact, the starting molecules give rise to the genetic molecules that effect all the other processes and to the uncannily similar chlorophyll and hemoglobin, which are the respective foundations of photosynthesis and respiration. There are, then, cycles within cycles within cycles, almost indefinitely. Moreover, the substances of all the natural sciences from physics to ecology are interconnected throughout. In order to understand the chemistry fully, one must be aware of the physical phenomena leading to chemical reactions. The properties and reactions of both inorganic and organic chemicals lead to a fuller understanding of the biochemistry and physiology at the cellular level, which in turn renders the physiology of the organism more comprehensible. Individual organisms make up populations that are the units composing communities that describe or characterize ecosystems subject to features of space and time.

Here, as always, each level of organization depends on those below, even as each new level manifests properties that could not have been predicted exclusively in terms of the predecessor science.

BASIC CHEMISTRY

PHYSICOCHEMISTRY

Chemical reactions depend on the physical configurations assumed by elements, molecules, and compounds. Elements are the unit of chemistry. Each element has properties unique to it, and it is the smallest unit manifesting those properties.

The properties depend on the number and arrangement of three particles: the proton and neutron of the atomic nucleus and the electron, which occurs outside the nucleus and participates in the chemical reactions characteristic of the element.

INORGANIC CHEMISTRY

Inorganic chemistry is that of all the elements—including pure carbon and carbon dioxide but excluding carbon in certain other reactions—usually associated with life.

Inorganic chemistry is essential to ecological phenomena. One of the most important categories of inorganic chemical reactions is oxidation and reduction, or "redox," reactions. This is the chemistry of electron transfers. The generalized redox formula is:

$$Ae^- + B \rightarrow Be^-$$

In this crucial family of reactions, there is always an electron donor, the reducing agent, and an electron acceptor, the oxidizing agent. In the course of the reaction, the reducing agent is oxidized—it gives up its electrons-and the oxidizing agent is reduced—it accepts those electrons. It is important to note that this is always a two-way reaction; there is no oxidation without a corresponding reduction, or reduction without oxidation. This rule of physical chemistry holds because electrons cannot be free but are always an integral part of at least one atom. A representative reaction is the bonding of hydrogen and oxygen to form water:

$$2H_2 + O_2 \rightarrow 2H_2O$$
$$(Ae^-)\ (B) \rightarrow (A[+]Be^-)$$

Both molecular hydrogen and molecular oxygen have an oxidation number of zero, because in each case electrons are being shared by the paired atoms. Hydrogen, a strong reducing agent, tends to release electrons, whereas oxygen, a strong oxidizing agent, tends to accept them. Significantly, water is polar because the electrons donated by the hydrogen are held more strongly by the oxygen. Thus, this neutral molecule is more positive in the vicinity of the hydrogen and more negative in the vicinity of the oxygen. This characteristic of water is one of the many that render it essential to life.

Water is the molecule usually considered to be the single-most important to life and its processes, but other simple inorganics and the elements are also essential, as will become increasingly apparent throughout this chapter. For example, certain of the much-maligned heavy metals are actually essential to life, if only in trace quantities as micronutrients.

ORGANIC CHEMISTRY

Organic chemistry is the chemistry of carbon, especially

in combination with hydrogen and oxygen. Carbon dioxide is something of a transitional compound. It, itself, is inorganic, but it is the source of the carbon backbone of all organic matter.

Carbon is unique among the elements in its capacity to seek chemical bonds for the four electrons of its outer electron shell. It can bond with one to four additional carbon atoms or with hydrogen, oxygen, or nitrogen. This ability leads to an astounding number and diversity among resultant molecules. Both factors are essential to the intricate chemistry comprising the features identified with life.

The simplest organic molecule is methane, carbon with single-bond connections with each of four hydrogens. Methane is a gas and not only the unit from which all biochemicals can ultimately be built but also one end product of catabolic processes. What may seem to be subtle differences in molecular structure can result in profound differences for life.

Carbon with single-bond connections with four chlorine atoms in place of the hydrogen is carbon tetrachloride, a poison. One hydrogen and three chlorines result in chloroform, a poison that humans have utilized as an anesthetic because initial toxicity takes the form of a deep induced sleep that prevents the sensation of pain.

Carbon can form single chains: the two-carbon chain is ethane when the three available electrons on each carbon are bonded with hydrogen. Replace one of those hydrogens with the hydroxyl group (-OH), and the result is ethanol, another toxic substance that humans utilize. Dilute ethanol is the alcohol of alcoholic beverages. Methanol, the single carbon with the hydroxyl group replacing one hydrogen, is also an alcohol, but even dilute methanol is highly toxic to humans.

Replace a hydrogen of ethane with the carboxyl group—an oxygen double bonded on the same carbon atom bonded to a hydroxyl, usually represented as:

$$\mathrm{COOH}\ \textit{or}\ \overset{\mathrm{O}}{\underset{\mathrm{C}}{\|}}\text{-O-H}$$

And the result is acetic acid, better known as vinegar. Acetic acid, the simplest of the organic acids, is only one of a series absolutely essential to life. Carbon chains can include carbon atoms double or triple bonded to other carbons, and once the chain is at least five carbons long, the two ends of the chain can connect. Six-carbon chains are hexanes or hexenes depending on internal bonding between the carbons. Both chains can actually include a split instead of simply being straight. The six-carbon ring compound with each carbon atom saturated with hydrogen (that is, every carbon bound to a carbon on each side with two single-bonded hydrogens) is cyclohexane.

Each of these simple six-carbon molecules has its own set of properties, with special reference to solubility in water and in polar and nonpolar liquids, density, and melting and boiling points. Cyclohexane is comparable to its straight- (or open-) chain counterparts, but it is nonpolar with slightly higher density and boiling point. It happens to be an excellent fuel for motor vehicles. More significantly to biochemistry, hexane is readily converted into the aromatic hydrocarbon benzene, the same six-carbon chain in a ring formation with six less hydrogens and the carbons alternating single and double bonds between them.

Benzene is the chemical unit of aromatics. This molecule and its aromatic five-carbon relatives contribute much to the variety requisite to cellular physiology. Each hydrogen bonded to the ring carbons can be replaced by groups such as hydroxyl or by another carbon, among other elements. A vast array of molecules can result, especially since two aromatic rings can be connected by a carbon or oxygen link or directly by sharing two carbons in common. All of these constructions of carbon play a role in living chemistry—biochemistry and physiology.

Essential to life, amino acids consist of at least two carbons in a chain. The first carbon bears a carboxyl group; the second, an amine ($-NH_2$), attached to the carbon by a single bond with the nitrogen. Most of the active amino acids contain only carbon, hydrogen, and oxygen, but two also contain sulfur.

All the hundreds of proteins, of which living organisms are constructed and which, in the form of enzymes, mediate every reaction of metabolism and catabolism, are constructed of linear arrangements of twenty amino acids in polypeptide chains, synthesized through chemical condensation (described below) within cells on internal structures known as ribosomes.

Biochemists identify proteins on the basis of the amino acids present; the order in which they are linked in polypeptides; and function, which depends on the elaborate three-dimensional folding that results from interactions among atomic components exposed to one another along the chains.

Enzymes, the specialized protein catalysts, by definition alter the rate at which specific biochemical reactions occur, without themselves undergoing permanent change. Thus, relatively small amounts of any given enzyme are required when called upon to serve physiological needs.

Some 2,000 enzymes are known to exist, and each catalyzes one specific reaction. Among these reactions are condensation, hydrolysis, transfers of carbons or of active radicals or other chemical moieties, and biological oxidation-reduction reactions.

Even more elaborate combinations of carbon configurations and an array of inorganic elements are found in molecules such as chlorophyll, hemoglobin (a protein), steroids, and vitamins.

THE BIOCHEMISTRY OF ECOLOGY

The overall photosynthetic reaction should look familiar; it is a redox reaction whereby the hydrogen of water is the reducing agent and the oxygen in carbon dioxide is the oxidizing agent:

$$6H_2O + 6CO_2 \rightarrow C_6H_{12}O_6 + 6O_2$$

$$Ae^- \; B \; Be^-A$$

Significantly, for biology and hence ecology, the addition of hydrogen creates energy-rich compounds such as sugars. Respiration, in effect the chemical reversal of photosynthesis, is also a redox reaction:

$$C_6H_{12}O_6 + O_2 \rightarrow CO_2 + H_2O + \text{energy}$$

$$Ae^- \; BA \; Be^-$$

It is a series of redox reactions in which carbon is being oxidized. The overall reaction is in essence a slow "burning" of sugar in which the energy of the hydrogen electrons is released; significantly, hydrogen-poor molecules are energy poor. In combination, photosynthesis and respiration are a shorthand for the generation and storage of cellular energy and its release for use in cellular processes.

On one scale, these processes are reflected in the ways in which biomass is weighed for purposes of analyzing structure or function in an ecosystem or its components. Wet weight is the total biomass, including all the associated water in the cells of organisms.

Dry weight measures the organic portion of the biomass, including all the inorganics incorporated in organic molecules. Ash is the inorganic component after the purely organic has been burned to CO_2 and H_2O, both of which will evaporate into the air in the heat of burning.

Carbon transfers, the buildup and breakdown of biological molecules composed of carbon chains and rings by the stepwise addition or elimination of single carbon atoms and bonded groups, are essential to the processes of life. Photosynthesis transforms the single carbon of carbon dioxide into the six-carbon compound glucose.

Glucose then enters the metabolic processes of the living plant in reactions found throughout the plant and animal kingdoms. (Similar, if not identical, reactions also occur among protista, bacteria, and fungi, based on the same overall biochemical themes.)

The first step for glucose is glycolysis, whereby each six-carbon sugar molecule is broken down (hydrolyzed) into two three-carbon molecules, pyruvate. The pyruvate is further broken down to form the two-carbon group, acetyl.

It is this group in association with a specific coenzyme that enters the Krebs or citric acid cycle, the centre of respiration in both plants and animals. In essence, this cycle constantly moves from two-to four-to six-carbon chains and back again along a precise chemical pathway. In the process, energy-rich molecules (ATP, adenosine triphosphate) are formed with the donation of electrons from hydrogen. Each time a carbon is removed from a molecule, it is as carbon dioxide. The associated hydrogen (electrons) is then carried out of the system by oxygen with the generation of water. This process provides the energy for the work of life, which is actually being carried out at the subcellular level. When energy is required, ATP reverts chemically to ADP (adenosine diphosphate), which can return to the Krebs cycle. Indeed, most, if not all, of these individual biochemical reactions are reversible. The reverse process, however, requires the presence of an enzyme other than the one catalyzing its counterpart reaction—another deceptively simple control mechanism in the chemical complex identified as life. Derivatives of these processes and their products are the chemical products of all the materials associated with life, including amino acids and the proteins and lipids, which include fats, oils, steroids, and waxes. Each process is mediated by the genes as they "switch" on and off. The genes are themselves specific organic chains composed deceptively simply of multiple nucleotides, each a combination of a five-carbon sugar, phosphate (PO_4^{2-}) and one of four organic bases. Each base consists of nothing more than carbon, oxygen, hydrogen, and nitrogen.

Another major reaction is condensation, the joining of monomers into larger "macromolecules" with the extraction of water. The generalized formula is again deceptively simple,

Monomer-H + HO-Monomer → Monomer-Monomer + H_2O One of the most important condensation reactions is that which occurs between two amino acids to form a dipeptide, tripeptide, and so on, into the polypeptide proteins:

```
   NH2   O            OH   NH2              NH2 O
    \   //             \ H /                 \  //
R -- C -- C     +      C -- C -- R   →   R -- C -- C     H  +  H2O
    /   \             //    \                /   \ /
   H     OH           O      H              H  HO  N
 Amino acid          Amino acid                     \ \
                                                     C -- C -- R
                                                   //
                                                   O
                                                Dipeptide
```

What has happened in this representation is that the hydroxyl group in the first amino acid and one hydrogen from the amine group of the second have been split off to form water. The carbon from the first then forms a bond with the nitrogen of the second. Note that each carbon still has all four bonding sites filled and no more than those four. (All four electrons are shared for each carbon.)

If one "straightens" the dipeptide molecule, its representation better reflects its functional components:

```
                    (amine)
                       NH2    O    H    H    OH
                        \     //   /    /    /
(specific
amino acid)  R -- C  --  C -- N  -- C -- C            (acid)
                  /                 /     \\
                 H                 R       O
                      Specific amino acid)
```

In the case of amino acids the condensation reactions continue, under the direction of the genetic material in the cell, to form specific polypeptide molecules, proteins. Specifically, the amine above can release an electron (one hydrogen) to combine with the hydroxyl group from a third amino acid. The acid (carboxyl) group above can release a hydroxyl group to combine with an amine hydrogen from a fourth amino acid. Thus, a polypeptide of four amino acids would be synthesized, with the release of an additional two molecules of water.

This condensation process continues to form the elaborate polymers known as proteins.

The chemical reversal of condensation is known as hydrolysis, the breakdown of macromolecules into their component monomers by the restoration of water:

Monomer-Monomer + $H_2O \rightarrow$ Monomer-H + OH

All cells hydrolyze macromolecules into their monomers and use the monomers for maintenance, growth, specific functions and reproduction. For example, a cell will break down proteins into amino acids in order to generate different proteins. Heterotrophs cannot make their . own macromolecules from "scratch," as can autotrophs, which is why herbivores must consume plants to survive. Similarly, carnivores cannot generate all their required macromolecules from scratch or directly from those generated by autotrophs. Carnivores must consume other heterotrophs that have already taken the biochemical steps carnivores cannot take. Omnivores can consume a variety of producers and consumers at more than one trophic level.

Alternative biochemical pathways accommodate to the sources of the organic materials ingested. Ecosystems, then, are composed of a community of organisms in which there are numerous cycles of closely interconnected pathways of producers, phagotrophic consumers, and saprotrophic consumers. Each cycle is connected to the abiota by green plants and by autotrophic and heterotrophic microorganisms.

The so-called scavengers—vultures, hyenas, sharks, and other (phagotrophic) consumers of carrion—can also be considered a part of the detritus circuit, but many insects and other organisms of comparable small size take some of the first steps to produce the molecules on which the communities of microorganisms feed.

Each species has a precise role in an ecosystem's trophic cycles—its niche. Some of those species can be replaced by others, depending on the specific ecosystem and its strategy over time. Other species may be uniquely capable of filling a niche in an ecosystem. If such a species is lost, that portion of the cycle is discontinued. Unless adaptations occur, the cycle will cease. If it is crucial to neighboring cycles, they will also cease. Some may adapt; others cannot. Ultimately, the whole ecosystem will adapt or be altered. If homeostasis is restored, the new community will continue to function as a recognizable unit, albeit at least subtly distinct from the original ecosystem. If the alterations exceed the system's ability to restore a homeostatic condition, the ecosystem will be lost.

How all these wondrous transformations proceed in keeping with the specific functions and changing needs of cells, tissues, organs, and the whole organism belongs in the black boxes of biochemistry, physiology, endocrinology, and a host of molecular biology disciplines well beyond the immediate scope of ecological principles. Suffice it to say that as genes are turned on and off, enzymatic reactions carry out the business at hand.

Enzymatic catalysis is the precise biological rating of biochemical reactions. Most reactions of metabolism and catabolism would never occur spontaneously or would proceed either infinitely slowly or so rapidly as to be ineffective or destructive in the biological system. The processes of life, therefore, depend on catalysis carried out by enzymes.

Both reactions whereby the products are at a lower energy state than the reactants (exergonic reactions) and those whereby the products are at a higher energy state than the reactants (endergonic reactions) are impeded by the need for a relatively large energy input—an energy "hump" that must be surmounted. Enzymes substantially reduce this hump by interacting to bring the reactants together in chemical "readiness" and then being released unchanged.

In biological processes, there are numerous specific interconnections of such reactions, with the energy released by an exergonic reaction frequently driving an endergonic one.

These processes occur at specific locations within the cytoplasm and subcellular structures, as required by the cell in its particular environment. For unicellular organisms, that environment is external—the abiota as it is constantly mediated by other members of the biota. For multicellular organisms, the cellular environment consists of other cells and any products they secrete in specific structural organizations building to the organism itself.

In biochemical processes, enzymatic catalysts function with cofactors or coenzymes. Cofactors are inorganic ions—often heavy metals—at the site on the protein molecule that actually serves the catalytic function. Coenzymes are nonproteinaceous organic molecules, often those characterized as vitamins or incorporating vitamins (as a "moiety") within the chemical structure. For example, the protein hemoglobin, which gives blood its red Colour, contains iron. It is the iron that chemically binds to and holds oxygen for distribution throughout the body and releases it as needed for the Krebs citric acid cycle. Other essential trace elements (also heavy metals important in toxicology) are magnesium in chlorophyll and copper in vitamin B_{12}.

Thus, when exploring biogeochemical cycles, the material cycling component of ecosystems, there are a number of specific elements and simple molecules of major interest. Primary among these elements is the carbon of carbon dioxide and of organic matter. Hydrogen and oxygen are also essential, both occurring in the biologically essential form of water, and the latter occurring not only in carbon dioxide and water but also as pure molecules: diatomic molecular oxygen (O_2), essential to respiration; and the unstable ozone (O_3), significant, on one hand, as a mutagenic agent and, on the other, as protection from another mutagenic agent, the ultraviolet wavelengths of solar radiation.

In considering these few elements and simple molecules and their seemingly countless reactions in only a few unmistakable basic patterns, the fundamental paradox emerges of the precisely structured and functional complexity of life based in deceptive simplicity. It is truly astounding.

Moreover, it can all be traced to the physical processes of the Big Bang and the formation of galaxies and solar systems. All of chemistry, inorganic and organic, is an elaboration of those physical processes. We and all of life are in a very real sense the stuff of the sun, which in turn is the chemical stuff of the galaxy and ultimately of the Big Bang. "What goes around comes around" is profoundly accurate in connecting life and the cosmos.

Chapter 5

Biogeochemical Cycles

PLANETARY ABIOTIC SOURCES

Holistically, in what appears to be a recurring strategy, abiotic conditions, notably including the substrate represented by geological features on the macro-and microscale, make it possible for life to arise and determine what life forms can become established. These "pioneering" organisms interact with the abiotic substrate, altering it and making way for additional or different organisms. The process may stabilize at a given stage for a measurable interval of sufficient duration to be recognized as an ecosystem, defined by its unique community.

It is instructive to examine planet Earth in terms of the three major components of the biosphere: the atmosphere, the hydrosphere, and the lithosphere. Together, they are the source of all the chemical components of life, and each has a place in every biogeochemical cycle.

ATMOSPHERE

Essential to the maintenance of life is the earth's atmosphere. This gaseous envelope is as important to the individual organism as it is for the global biosphere. Today's atmosphere is an oxidizing one. Early in the evolution of the planet, molecular hydrogen predominated for a reducing atmosphere. Life, then, first originated in a reducing atmosphere and actually created the oxidizing atmosphere. In order to survive, organisms then had to adapt or evolve accordingly. One could argue that life itself has been a major perturbation of the global environment. One of its first influences was the poisoning of the atmosphere with its wastes. Fresh perspective on contemporary environmental crises is provided through recognition that life then proceeded to evolve in accommodation of the toxic atmosphere it generated. The biosphere we hope to preserve and protect is still being maintained by the planetary biota, including human populations.

Today, as a result of life's historical influence on the biosphere, slow oxidation in "air" is occurring constantly. Iron and other metals rust, or, more precisely, oxidize. Fire occurs with some frequency. When organic material burns, it is being rapidly oxidized. Complete burning, which occurs in the presence of ample oxygen, results in the generation of CO_2, H_2O, and energy (the heat and light of fire). If any heavy metals are present, they will be restored to the elemental form found in the ash.

There are a number of logical approaches to the earth's atmosphere. It has thermal, electrical, and compositional properties. Electrically, the lower sixty or so miles from the planetary surface constitute the neutrosphere, electrically neutral and hence relatively unreactive—leaving the reactions to the chemistry and biology of the hydrosphere and lithosphere. Above the neutrosphere is the ionosphere.

The ions that give this region its name would be extremely reactive were they not separated by decreasing concentrations, a reflection of the tendency toward entropy as these entities escape the atmosphere for outer space.

The temperature decreases with distance from the planetary surface. Temperatures are highest and the most dynamic below the stratosphere, especially in the lowest five to thirteen miles where weather is occurring. A thermal component can be measured as high as 100 miles from the surface, in the thermosphere. Both the concentration of ions and temperatures ordinarily measured as Centigrade or Fahrenheit continue to dissipate with distance from the planetary surface until interplanetary space of the solar system is attained.

It is illuminating to follow the chemical composition of the atmosphere from those outer reaches down toward the surface. The outermost region, from 1,500 miles in altitude and beyond, is the protonosphere, essentially molecular hydrogen. As altitude decreases, hydrogen gives way to helium, which gives way to molecular oxygen.

Oxygen gives way to an oxygennitrogen mixture at an altitude of about 75 miles. Almost completely coextensive with the stratosphere is the ozonosphere, some 6 to 50 miles from the surface. Here the oxygen-nitrogen mix gives way to the unstable ozone molecule composed of three, rather than two, oxygen atoms.

The chemical mix from the stratosphere to the atmosphere is dynamic. The lower sixty miles are the homogeneous mix familiarly known as "air," where the winds are active. The ecosphere, the portion of the atmosphere sustaining life, is only the last two miles or so.

The ecosphere is composed mostly of molecular nitrogen, just over three quarters of the chemical content of dry air measured by volume or by weight. Nitrogen, significantly enough, is relatively stable. Molecular oxygen is just over a fifth of the content, also by volume or weight. A greater concentration of oxygen would result in an explosive atmosphere, whereas a lesser concentration would be suffocating to the predominant aerobic organisms of earth.

The noble (nonreactive) gas argon is the next-most-concentrated component, at just about 1%. Carbon dioxide is present at less than 0.5% by volume. Hydrogen, ammonia gas, and ozone are present in volumes of 0.0001% to 0.00001%. Since each of these gases is toxic to life or otherwise hazardous, albeit essential to biochemical processes, these minute percentages are significant to life and ecology for more than one reason.

HYDROSPHERE

The hydrosphere may be what sets Earth apart from any other planet in the solar system otherwise theoretically capable of supporting life. Neither Venus nor Mars currently possesses a hydrosphere, a factor almost certainly precluding life on either planet. Venus is also too hot and Mars too cold for most forms of life, but temperature is one of the reasons earth does have a hydrosphere.

The hydrosphere at temperature regimes from the equator to the poles consists of water in three physical states: solid, liquid, and gaseous. The hydrosphere will be considered in greater detail as representative of hydrogeochemical cycles.

LITHOSPHERE

The lithosphere is the earth itself. This sphere, the subject of geology, is the substrate for both aquatic and terrestrial life. Like the atmosphere, the lithosphere is composed of distinct layers, each having a unique role in ecological processes.

The average radius of the planet is 4,000 miles. On the top of the solid matter of the planet is the relatively fine material known as soil. This is a complex and extremely variable sediment in terms of texture, inorganic content, and organic content. Soil is usually defined by its origins or its level of fertility.

The highly heterogeneous material is divided into two categories: topsoil and subsoil. The former is a mixture of highly decomposed mineral constituents and organic material known as humus. The underlying subsoil provides nutrients and moisture for deeply rooted plants. Beneath the subsoil at varying depths is bedrock, the outermost portion of the mantle, the lithosphere itself.

This sixty-mile crust consists of some ten miles of bedrock. The lithosphere is 99% silicates, oxides, carbonates, and phosphates, indirectly available to the biota in that physical processes are required to raise them to the surface where they are exposed to weathering, erosion, and other mechanisms breaking them down into formations into which life can intrude. Below this crust is the barysphere, the heavier interior of the mantle. The outer core of this region is molten rock, the source of volcanoes and of the heat of thermal waters. The inner core is also hot rock, but internal pressure is so intense that it remains solidified.

It is the bedrock that is of the most immediate significance to ecology. Nearly half of bedrock is oxygen, chemically bound in a variety of substances. More than a quarter of bedrock is silicon, relatively inactive in biological terms. Metals, such as aluminum, iron, calcium, sodium, potassium, and magnesium, are each present in concentrations approaching 10%.

Other metals and hydrogen are present in amounts ranging from 0. 04% to 0.1%. Among these are elements essential to life, notably carbon, sulfur, and phosphorus. Significantly, none of the three planetary sources of material contains that material in the form of relative concentrations found in the biota.

To maintain its distinction from the abiota, life obviously directs substantial energy to building and maintaining its characteristic chemical organization.

WATER AND THE HYDROSPHERE

Water and the hydrosphere can be considered a biogeochemical cycle, with water representing the nutrient of interest. But water is also a major carrier and reservoir of all other nutrients. As the liquid dihydrogen oxide, water is truly a unique substance and one particularly suited to supporting life. Physically, it is transparent to incoming sunlight but absorbs reemitted infrared to serve as an atmospheric temperature control.

Through the physical process of evaporation and the biological process of transpiration, water also holds terrestrial temperatures below lethal limits for plants and animals. The atmosphere can retain more water as its temperature increases. This process stabilizes ambient temperature in the event of a sudden input of heat—as regularly occurs, for example, with sunrise.

THE CHEMISTRY AND BIOLOGY OF WATER

When water is formed by the reduction of oxygen and oxidation of hydrogen, the electrons are shared by the two elements composing the molecule, but oxygen holds them more strongly than does the hydrogen. Hence, although water has an oxidation number of zero, the two hydrogens of water tend to have a positive charge, while the oxygen tends to have a negative charge. Water is therefore polar—one of the properties making this molecule so important to life. The polarity of water makes it a good solvent for inorganics but not for organics. Hence, inorganic and simple organic nutrients dissolve in water, but organisms and their principal molecules do not. It is notable that water does, however, constitute up to 90% of the "chemical" content of organisms.

Water itself readily dissociates to yield $2H^+$ and O^{2-}or, more accurately, H_3O^+ (hydronium ion) and OH^-(hydroxyl ion). These ions are highly reactive with each other and with other ions in aqueous solution. This property of water is essential to its ecological role.

When the hydronium and hydroxyl ions are present in equivalent amounts, the resultant solution (environment) is neutral, designated as pH 7.0. Sometimes, however, reactions with ions in solution result in acidic conditions, where more hydroxyl than hydronium ions have entered chemical combinations that remove them from solution, leaving an excess of hydronium ions, designated as pH levels less than 7.0.

The reverse can also occur to leave an excess of hydroxyl ions in solution, a basic condition, designated as pH levels greater than 7.0. What is being measured is the concentration of hydronium ions, expressed as a negative coefficient. That is, neutrality is a concentration of 10^{-7}. A pH of 5, for example, is a 10^{-5} concentration, while pH 8 is 10^{-8}. There are more hydronium ions at the lesser pH than at the higher pH.

Each organism and many biochemical reactions depend on the pH of the immediate environment. Moreover, the availability of certain nutrients and toxic chemicals to the biota is closely related to pH.

Thus, the health of an ecosystem can be affected positively and negatively by alterations in pH. Such effects can rise to the level of perturbation (too little or too much nutrient, altering toxicity of substances present), which may be natural or anthropogenic.

THE HYDROGEOLOGY OF ECOLOGY

As a preliminary to biogeochemical cycles, hydrogeology serves two purposes: first, as a carrier of ecologically important chemicals; second, as a model for biogeochemical cycles. Since the processes of interest are cycles, it does not matter where the starting point of analysis is set. Water is a major constituent of the atmosphere and is found on and in the lithosphere, but water is most prevalent as oceans. The oceans contain literally billions and billions of grams of water—nearly a billion and a half billion billions of grams.

To gain perspective, a pound weighs some 454 grams, and a billion billion is more accessible when converted to exponents, sometimes known as scientific notation: 1×10^{18}. A relatively small portion of the vast reservoir of water enters the atmosphere through evaporation, some 319×10^{18} grams. Some, but not all, of this evaporated water will return to the oceans as precipitation, 283×10^{18} grams.

The rest remains in the air as water Vapour, with some of the water molecules forming aggregates that are sufficiently concentrated to be visible: clouds. Containing less than 40×10^{18} grams of water, clouds are moved by net wind movement in patterns around the globe. When conditions are suitable, the water in clouds returns to land formations as precipitation, nearly 100×10^{18} grams.

Some of that water runs off the land surface, to enter bodies of surface water, which ultimately make their way back to the oceans. Some of the precipitated water enters the ground and finds its way into underground waters, a system of streams, ponds, and lakes known collectively as groundwater. Together surface runoff and groundwater represent nearly 40×10^{18} grams. The remaining water percolates through the soil and enters still deeper underground systems, including cracks and pockets in bedrock.

Over time, earth's sedimentary rocks have incorporated much water, physically and chemically, representing the second-largest reservoir of water on earth with a content of just over $200{,}000 \times 10^{18}$ grams.

What is the source of the additional water entering precipitation over land? Nearly 60 × 10^{18} grams come from the physical process of evaporation from land and the biological process of transpiration from the leaves of plants. Water enters vegetation through the root systems and is carried throughout the body of each organism. With it comes the inorganic nutrients needed to initiate the grazing cycle. But were that water retained in each plant, it would soon become bloated. Specialized cells in the leaves open and close to control the release of water into the atmosphere.

All those processes in equilibria with each other make up the water cycle, quite clearly composed of chemical, biological, and geological reactions or processes. Additional water movement is associated with vulcanism, is bound up in the polar ice caps, and is lost to outer space.

EPRESENTATIVE BIOGEOCHEMICAL CYCLES

With the water cycle as a model, a generic biogeochemical cycle can be described. There are three major components, which, significantly enough, represent three major pieces of the generic ecosystem: available inorganic nutrients, living organisms, and organic detritus. Again, material is cycling, so analysis can commence at any point without warping logic.

This time it is instructive to begin outside the immediately active abiotic portion of the cycle, with indirectly available inorganics. This material is, in fact, the lithosphere itself, more specifically, the bedrock and even more inaccessible reaches of the earth's interior. This indirectly available matter, which is inorganic, undergoes weathering, erosion, and biological fixation in a variety of forms ranging from the physical action of deep roots to the biochemical reactions of microorganisms. This array of actions renders the inorganics available to the grazing cycle. They become biologically fixed through the process of assimilation and production through the trophic levels of each ecosystem.

Respiration, leaching, and extraction return some of the nutrients from living organisms to the available inorganic nutrient pool. Much of the material incorporated into the living mass enters the organic detritus. Some of the detritus returns to the biota through feeding by detrivores.

Much of the organic material is leached, burned, or decomposed, with a release of inorganics into the pool of available inorganics and returns again to living organisms. The chemical components of some organics become indirectly available and are restored to the inorganic cycling pool through burning, leaching, or decomposition and to the organic detritus through erosion.

Some of the organics form the peat, coal, and oil that are stored indefinitely in the lithosphere. The inorganic nutrients incorporated in these organics are restored to the inorganic pool by burning. Similarly, some of the available inorganics are returned to indirectly available status through precipitation and sedimentation. They are returning to the major reservoirs: the atmosphere, portions of the hydrosphere, and the lithosphere.

Biogeochemical cycles can be represented in two lesser cycles: reservoirs and pools. As the designations imply, the difference is a quantitative one with some relationship to availability as well as to speed of movement through the environmental components. Generally, a reservoir pool is large, slow moving, and generally abiotic.

A cycling or exchange pool is smaller, more active, and involves transactions between organisms and their immediate environment, sometimes considered the microenvironment. It is through all these processes—physical and chemical as well as biological—that the 30 or 40 elements known to be required by organisms are extracted from more than 100 known chemical elements, driven by the one-way traffic in energy flowing from the sun to outer space.

CARBON

The element carbon represents 0.03% of the atmosphere as the inorganic gas carbon dioxide. That gas is in constant exchange between the atmosphere and the ocean. In this and other aquatic systems, carbon's representative inorganic molecules are also in chemical equilibrium:

$$CO_2 + H_2O \rightarrow H_2CO_3 \rightarrow H^+ + HCO_3^- + CO_3^{2+}$$

This shorthand equilibrium, by which carbon dioxide can become a carbonate ion with intermediate stops at carbonic acid and a bicarbonate ion, results in a buffering capacity in bodies of water. Either an excess of acid or an excess of alkalinity will be chemically assimilated by the equilibrium to restore the system to the pH existing prior to the perturbation.

Hydronium ions will react with cations in bases to remove them from the system, while the bicarbonate or carbonate ions will react with anions in acids to remove them. In a specific ecosystem, carbon will be found as carbon dioxide, carbonates, organic detritus, and most important, living tissue. The biogeochemical cycle of carbon is already familiar.

The major reservoirs are the atmosphere and the hydrosphere with cycling pools at or near the earth's surface and at the surface levels of water where light can penetrate and photosynthesis can occur. A major organic reservoir of carbon is represented by peat, coal, and oil, each of which must undergo the rapid oxidation called burning before it can reenter the atmosphere and hydrosphere and the associated exchange pools. Carbon is also in the lithosphere in the form of limestone and dissolved marine humus. All of these forms are biogenic.

In summation, most of the planetary carbon is in the abiota, primarily the hydrosphere and lithosphere. Only a small amount at any given instant can be found in either the atmosphere or the biota.

OXYGEN

In contrast to carbon, oxygen represents a little more than 20% of the atmosphere. It is also present as water. Its exchange pools are the water and the atmosphere, where it is present both as carbon dioxide and as molecular oxygen. It is also a major component of a variety of inorganic compounds, designated by the chemical suffixes "-ite" or "-ate," as in nitrites and nitrates, phosphites and phosphates, carbonates, and bicarbonates. Acids and bases, organic as well as inorganic, also contain oxygen. Biologically, oxygen can be considered a by-product or waste product of photosynthesis or as one of the three major chemical building blocks of protoplasm. The role of oxygen in respiration is one of oxidation: It accepts the electrons of hydrogen and removes them from the Krebs citric acid cycle.

The biogeochemical cycle of oxygen is as familiar as the carbon cycle, of which this element is a part. It is as nearly ubiquitous throughout the biosphere as any element can be. Less familiar forms are found chemically bound up in rock formations, including granite, marble, and quartz.

NITROGEN

Secondary only to the carbon, hydrogen, and oxygen in organic molecules is nitrogen. Molecular nitrogen, N_2, represents nearly 80% of earth's atmosphere, but it is essentially useless to the biota. Molecular nitrogen can neither be photosynthesized nor respired. It can be absorbed (principally a physical process) but not assimilated (a biological process). Yet it is prevalent in the soil as organic matter. Somehow it is drawn into the grazing and detritus cycles and converted into amino acids to generate proteins. Nitrogen is also a part of the bases encoding the directions for all life processes mediated by the genetic molecules DNA and RNA. Nitrogen is even a component of chlorophyll, used by the biota to fix carbon dioxide and light.

Nitrogen is involved in a biogeochemical cycle, but this time a logical starting point can be selected. It is the biological fixation of molecular nitrogen. In fact, the intricacies of interactions in the nitrogen cycle are a particularly lucid model for all the remaining nutrient cycles. Each movement in the cycle is actually one step, mediated by a specific organism when not subject to purely physical and chemical alterations.

Molecular nitrogen in the atmosphere is biologically fixed by genera of nitrogen-fixing bacteria. Their habitat is found within the roots of certain plants, the legumes. Here, in a symbiotic relationship, the bacteria convert nitrogen to nitrate, an anion of nitrogen combined with three oxygens.

The chemical compounds known as nitrates are soluble in water. They are carried into the soil where they are absorbed and assimilated by green plants at large and thus enter the two trophic cycles as organic nitrogen, frequently in amine form. Nitrates can also be assimilated by additional genera of bacteria, the denitrifiers. They convert nitrates to nitrites (one less oxygen atom in the anion) and nitrites to molecular nitrogen, which is restored to the atmosphere.

Animals excrete urea and other organics, which are converted by soil bacteria to ammonium compounds (containing the cation NH_4^+), which nitrifying bacteria convert to nitrites (*Nitrosomonas* spp.) and to nitrates (*Nitrobacter* spp.), which can enter the soil for cycling through the grazing cycle. The nitrates can also be denitrified by additional bacterial taxons. As both plants and animals die, their nitrogenous biochemicals are acted upon by putrifying bacteria.

Each biochemical step (some of which are the familiar redox reactions) in the "bio" portion of the nitrogen cycle is also an energy transfer. There is an energy barrier, surmounted by enzymatic catalysts unique to nitrogenfixing bacteria, in the conversion of molecular nitrogen to nitrate. Each step from the nitrate through nitrite, ammonia, and amino acids to protoplasm requires external energy, because each is chemically more likely to proceed in the opposite direction. That energy comes from the sunlight or from the breaking of chemical bonds in organic matter. It is operating against entropy.

HEAVY METALS

Phosphorus is also essential to life. Organic phosphates are the immediate source of energy in the cell. ATP holds the energy ready for utilization and is converted to ADP as the energy is taken. ADP enters the Krebs cycle and is "reenergized" with a new phosphate ion. Phosphorus, chemically a metal, is found in sediments more than in the atmosphere or hydrosphere, as is the case with other metals essential to life. Iron is an active component of hemoglobin; magnesium, of chlorophyll; and copper, of vitamin B12, a coenzyme. Magnesium is involved in protein synthesis in plants.

It is also found in the bone of vertebrates and is essential to enzymatic reactions mediating the transfer of phosphates. Together phosphorus with calcium actually constitute 70% of the total inorganic component of animal cells. In all organisms, phosphorus is involved in energy transfers and is a structural component of DNA and RNA. The calcium-phosphorus ratio is crucial for the utilization of certain vitamins and/or other nutrients.

In plants, calcium is involved in the association of cells and in root growth. In animals, calcium is involved in acid—base relationships and in blood clotting, the contraction and relaxation of heart muscle, and the control of fluid passage across cell membranes. The metal is a component of exoskeletons, shells, and skeletons.

The bio portion of all these biogeochemical cycles is reflected in the biomass, each in a way uniquely informative to the ecologist who would follow the fate of the chemicals involved. The relatively minute contribution of inorganics is informative in another way. Because so little is needed of each of these metals, too much in the ecosystem can be at least as harmful to the biota as too little.

Thus, these trace elements that support life are also known as *heavy metals,* a term carrying a connotation of toxicity, acute and chronic. Whenever dealing with chemical substances, a number of questions are crucial not only to the ecologist but also to the environmental professional who seeks answers to questions associated with regulation and litigation.

IMPLICATIONS FOR ENVIRONMENTAL PROFESSIONALS

Chemistry is obviously as important to the environmental professional as it is to ecologists. Both will be asking the same kinds of questions about a chemical substance in an ecosystem. The first and most obvious question, the answer to which may be considerably less obvious, is, What is the chemical in chemical terms? This question can be broken down into chemical and physical components. The identity of the unknown material is first what elements are present and in what combinations. Each combination will have unique properties, causing its actual placement in the ecosystem. A chemical substance will often occur in a mixture with the materials of the environment. By definition, the atmosphere, water, and soil are all mixtures. If in water, the substance can be in suspension or solution or somewhere in between as a colloid, each of which will have different implications for the material's physical, chemical, and hence, biological fate.

If chemical reactions are in progress, the substance may be in chemical or biological equilibrium, which will affect its availability to the biota. The next series of inquiries, which depend on accurate answers to the foregoing, address where the material is. Is it in the abiota, or has it entered the biota? Its present location determines where it is going and by what routes and through what alterations.

These are qualitative and then quantitative questions. In brief, a substance will either be available or unavailable to the biota or specific organisms. Available substances are of immediate interest. Those that are unavailable are significant in terms of the mechanisms by which they may become available.

By its nature, its location in the ecosystem, its amount and concentration, and its availability, a substance may be quantitatively insufficient to have any effect on the system. As its concentration or its absolute amount increases, a progression of effects in organisms may be detected. The substance may be a nutrient in micro-or macroquantities. As such, an excess of the substance may represent nutrient enrichment that can be a perturbation of the system as a whole or in terms of a portion of the system of particular interest to the observer. Whether a nutrient or not, increasing amount or concentration may begin to cause toxic responses, which are described as subchronic, chronic, acute, and lethal as the crucial quantitative parameter increases.

These effects may be felt only in one part of an ecosystem, but they can proceed to injure additional parts or to affect the ecosystem progressively but indirectly. These phenomena are illustrated in the timely matter of acid deposition.

ECOLOGICAL EFFECTS OF ACID DEPOSITION

THE CHEMICAL PATHWAYS

The major industrial precursors of anthropogenic acid deposition are sulfur dioxide and nitrogen oxides. These ions readily react in aerosols to generate sulfuric and nitric acid, respectively. The relevant chemistry of sulfur dioxide is better known than that of nitrogen oxides.

Natural rain, water without these particular contaminants, is slightly acidic at pH 5.6 and subject to control through the dissociation of carbon dioxide. Natural waters can have a pH as low as 5, suggesting the presence of naturally occurring acidification.

Natural sources include sulfur in the lithosphere, released by volcanic action, sea salt aerosols, forest fires, and microbial decomposition. Nitrogen in the atmosphere is converted to nitrogen oxides by lightning. That compounds of both sulfur and nitrogen can be biogenic should be abundantly clear from the foregoing descriptions of the chemistry underlying ecology. Transportation of these acids and their precursors is a physical phenomenon.

They are picked up and transported from one locale to others by the prevailing winds. In Europe, studies conducted in the 1970s indicated a residence time of up to four days that carried the substances more than 600 miles from their sources. Seasonal cycles were detected. In the United States, transportation is generally out of the west, eastward and northward. Deposition then occurs in precipitation in all its forms. Today, precursors to acid deposition are identified as major sources of air pollution.

In the case of sulfur dioxide, 80% of that released into the atmosphere is attributed to human activity, 100% in some regions. Of that 80% to 100%, 85% is attributed to fossil fuel combustion. Nitrogen oxides also come from combustion, the most notable source being motor vehicles.

The earliest reference to anthropogenic acid deposition came from England in 1727, followed by observations of acidification associated with a Swedish smelter in operation for 500 years. The atmosphere was reportedly so corrosive that no herbs survived in the vicinity of the smelter. Widespread recognition of the effects of reduced pH did not come until the 1960s. Following an American synopsis of deposition chemistry in the mid1950s, a continuous record of the phenomenon was initiated in 1963.

Historic pH levels are necessarily inferred, not measured. Researchers disagree profoundly on the matter of trends over the last 25 years, but indirect evidence represented by acidity in Arctic ice caps suggests an increase in acidity.

ACID DEPOSITION IN ECOSYSTEMS

While there are economic and social effects of acid deposition on human artifacts, the effects of acid in ecosystems are of considerable interest as well. In forests and crops, decline and dieback of trees have been reported, possibly exacerbated by drought or dry summers. Some detrimental effects are directly the result of altered leaching of cations and other chemical entities caused by a lowered pH.

Aquatic systems are to some extent buffered. But they, too, can be adversely affected. Changes in pH can affect communities of bacteria, algae, invertebrates, and fish, altering species composition and productivity, reducing numbers, and impairing reproduction and decomposition.

Elements of the progression of effects have been reported for public consumption from field experiments. The subject lake had a natural pH of 6.5. Four years into the research, pH was reduced to 5.9, and relatively obvious effects were detected in the biota. While phytoplankton remained abundant, the mix of species had altered. Freshwater shrimp had nearly disappeared, and reproduction in fathead minnows was failing. In the next two years, pH was down to 5.6. Mats of slimy algae were forming. Crayfish had become unusually vulnerable to parasites and were suffering anatomical damage. The decline in fathead minnows led to a temporary boom in another species. Ironically, the favored species was preferred by trout, which presumably benefited from the ecological alteration. By the seventh year, pH was down to 5.0. While there was no mention of the condition of the producers, the crayfish, leeches, and mayflies had all disappeared. Reproduction among the now-emaciated lake trout ceased. Thus, in seven years, a result upon which regulators could readily base governmental controls—assuming the source of acid to be a recognizable legal entity—had occurred: the death of a population of organisms of immediate human interest. Significantly, more subtle changes earlier in the progression could have supported political action to prevent this ultimate harm.

The question is the matter of political justifications and ramifications. Do we want a healthful or a healthy environment—and is there a meaningful difference? Even at such a profound level of perturbation, it was observed that at least partial recovery is rapid. Restoration of the ecosystem, however, is estimated to require hundreds of years. Clearly, the questions to be raised begin—but only begin—with ecological realities.

Chapter 6

Cosmology of Planet Earth

FROM THE BIG BANG TO THE SOLAR SYSTEM

Science currently cautions us that nothing throughout the universe is static. Everything, including the universe itself, is dynamic. Not only are all natural phenomena in constant flux, but they manifest direction. In keeping with entropy, "time's arrow" flows from the past through the ephemeral present into the future, never the reverse.

Moreover, strategies can be discerned. Without need to call on a sentient directing agent for purposes of scientific analysis, events reveal a destination. Everything from the universe through the unicellular organism has a beginning, a series of developmental stages and, presumably, an end. The Big Bang represents the beginning of the universe and, significantly, all its contents. Stars are born, evolve, and die, as do solar systems. Planet Earth reveals a strategy that has led to the establishment of a biosphere.

The biosphere is made up of interconnected ecosystems, each representing a substrate or habitat into which living organisms invade and then proceed to succeed one another. As the arrays of organisms proceed through successional stages, they alter the abiota even as it influences them. A climax community may result and then alter with the passage of long intervals of time in something like an aging process leading inevitably to senescence.

Hence, any view of a natural system, such as an ecosystem or some microcosm of or within biological entities, is truly no more than a snapshot. It is a picture of things as they were at the moment of observation—nothing more and, frequently, substantially less.

COSMOLOGY AND THE BIG BANG

Why, one might legitimately ask, would the greenhouse effect be in the same chapter as cosmology and the Big Bang? What possible connection can there be? One might also be tempted to ask what subatomic physics has to do with studies of the history of the universe.

But, as has been noted, there is a kind of hierarchy of sciences of which ecology is an integral part. Moreover, "every atom in our bodies, every atom we touch and breathe, was created long ago deep inside a long-forgotten star. We are literally made of the dust of stars."

Physicists and cosmologists currently assert that the only time that something came out of nothing occurred in the Big Bang, the theoretical creation of the universe. All forms of energy and all of matter from the hydrogen atom to the most complex biological molecule can be traced to that "singular" event.

When energy became matter, it was through the fusion of hydrogen nuclei (protons) to form helium. Until recently, those two elements in combination were thought to compose more than 99% of the material in the universe. Now there is reason to believe that "dark matter" is more abundant than the brightly visible stars and galaxies.

It is only in recent years that projections backward in time have been amenable to scientific probing. With cosmology, physicists have joined with philosophers and theologians to ponder origins and first causes. Competing theories have constantly emerged, but it was not until the middle of this century that theories gained sufficient scientific power and experimental techniques enough sensitivity to provide meaningful data.

In very recent times, the Big Bang, an extraordinary explosion some 20 billion years ago, has become the respectable *theory* of the origin of the universe and its structure and "Behaviour."

A variety of empirical revelations are consistent with the theory, although many mysteries remain unsolved, including that of dark matter. In keeping with the practice of science, the most promising contemporary contributions have resulted from multiple contributions. Significantly enough, the theory is maintained and advanced by empirical and experimental research conducted by physicists studying relationships between the energy and material resulting from the creation of the universe and astronomers studying the resultant large-scale structures, including that of the universe as a whole. These ambitious research efforts have called upon particle theorists and cosmologists to join talents.

Because the importance of the fourth dimension is a particular emphasis in ecology and with constant reference to the arrow of time in flight through entropy, the relationship of time to cosmology and to the Big Bang theory is of special interest. For environmental professionals, three different arrows of time must be taken into account.

From the largest to the most intimately associated with the human environment, they are the cosmological, the thermodynamic, and the psychological. Just how far one can connect ecology to the Big Bang requires an understanding of the nature of stars, the manufacturing facilities of the universe.

EVOLUTION OF STARS

Ecology, as we have seen, depends in large part on chemistry. Chemistry, in turn, is the study of the elements and their physical and chemical properties. But what is the source of the elements? The chemical elements are the stuff of which stars are made. Thus, chemistry is inextricably linked not only to the physics holding atoms together but also to the stars that find their origins in the energy and matter released as a result of the Big Bang.

The contemporary understanding of the life history of stars begins with space globules. Space is not a complete vacuum. Individual atoms are scattered throughout the great vastness of the universe. At some locations, there are more than the random average of atoms. At these locations, there is a concomitant increase in gravity. Increased gravity draws still more atoms with progressively intense concentration. Eventually, the globule becomes unstable, unable to support its own weight. With the pressure of "trillions upon trillions of tons of gas pressing inward from all sides," the mass contracts with increasingly greater pressures and densities and correspondingly increased temperatures at the centre. The gases begin to glow and radiation filters outward in the dark red portion of the electromagnetic spectrum. This is a protostar.

Like its predecessor globule, the protostar is unstable because it cannot support the weight of its outer layers. The inner contraction continues with still greater increases in pressure and temperature. When the central temperature reaches 10 million degrees, hydrogen "burning" is ignited. This is not combustion (oxidation) but the fusion of colliding protons, the nuclei of hydrogen. With each fusion, a helium nucleus is formed, with a loss of matter.

Einstein's well-known formula $e = mc^2$ applies, and a tremendous amount of energy is released from the lost matter, a thermonuclear reaction. Because the helium product is lighter than the hydrogen reactants, the contraction comes to a halt, and a star has been formed.

The newly formed star will continue to burn hydrogen for billions of years before becoming unstable again. Now rich in helium, the core shrinks with the inevitable temperature increases. At this stage in the star's life history, the hydrogen between the core and the surface ignites in the thermonuclear burning. As its outer layers are pushed outward, the star swells, with a contracting core and thin layer of burning hydrogen.

When the core reaches 100 million degrees, the helium nuclei fuse with the formation of carbon and oxygen from this thermonuclear reaction. With helium burning in the core and the shell of burning hydrogen, the star's volume increases a billionfold. As a result, the atoms of the outermost layers move farther and farther apart, with a concomitant reduction in density, pressure, and temperature. Relatively speaking, the star's surface is now big and cool. It will now evolve to a red giant with a surface temperature of 3,000 degrees.

When helium burning is depleted, the core becomes unstable yet again. Again it will contract. This time the outer helium will ignite and the inner mix of carbon and oxygen becomes inert for lack of sufficient mass to ignite any further thermonuclear reactions.

This configuration, like those before it, is unstable, but this time the star pulsates in a series of expansions and contractions. Each pulse of thousands of years represents alternating cooling and contraction, the latter reigniting internal fires. In the end, the outer layers, however, will separate completely to leave a stellar corpse behind. The "nebula" is short-lived—in only a matter of 50,000 years, the expanding envelope will disperse and disappear. The star itself will shrink as its warmth radiates into space until it becomes the size of the earth, a white dwarf.

THE SOLAR SYSTEM

The Milky Way galaxy was formed some 7 billion years ago, with our own solar system taking form about 4.5 billion years ago. The centre of our solar system is a hydrogen-burning star. A second-or third-generation star, some 2% of the sun's matter consists of elements such as carbon and oxygen. The sun is expected to continue burning hydrogen for another 5 billion years before becoming unstable.

Eventually, its surface will cool to a few thousand degrees with its white-hot gases and evolve into a red giant. In the process, it will vaporize the planet Mercury, sweep away the carbon dioxide atmosphere of Venus, and devastate Earth with boiling oceans and melting rocks. Eventually, our sun will become a white dwarf, about the size of Earth.

For all we know thus far in our intellectual history, Earth is unique among planets as an abode for life. While the history of this planet within its solar system can be described in terms of coincidence, random events, and statistical probabilities, a pattern can be discerned. A variety of scientists infer the history of planetary formation and geological evolution from an incomplete record left by the events, including corresponding events in the ephemeral present. Scientists have come to conclude that the Earth has been in constant evolution, with each step depending on the lengthy past.

This dependence attaches to the origin and evolution of life. As far back as the Big Bang, events in the universe as well as on Earth itself have placed important restrictions on life's nature and its evolution. The lessons of paleontology, including the remnants of paleoecology, can also be turned to prediction. The accuracy of such predictions varies wildly. Sometimes it seems as though we can only predict into some vague middle time but neither the immediate nor the distant future, a function actually of the chaos side of natural events.

GEOLOGICAL EVOLUTION

Planet Earth is currently believed to be about 4.5 billion years old, measured with the completion of the lithosphere. After about 500 million years of geological evolution, life occurred in the oceans. As is the case with every good ecosystem, life invaded while the abiota was still undergoing its own transformations. From that point on, the biosphere—lithosphere, hydroshere, and atmosphere—continued to evolve, but now it evolved in concert with the biota, each influencing changes in the other in a discernible planetary pattern.

ORIGINS AND STRATEGIES OF LIFE

Almost all of life is distinguished from nonlife by the production of cytoplasm, the highly organized chemical stuff of life separated from the nonliving environment by a membrane, itself an intricately functioning component of the living cell. *All* of life shares the molecule DNA or, at least, RNA.

Even viruses, entities on the very border separating biology from chemistry, while lacking cytoplasm or any cellular structure of their own beyond a simple protein coat, all possess either DNA or RNA. For those laboring under the misapprehension that life is fragile, it is illuminating to view life as a planetary membrane—the "toughest... imaginable in the universe, opaque to probability, impermeable to death." Life is astonishing in its uniformity as well as in its diversity. There is a high probability that all of us derive from a singe parent cell. And "[i]t is from the progeny of this parent cell that we take our looks; we still share genes around, and the resemblance of the enzymes of grasses to those of whales is a family resemblance."

One of the earliest innovations of Lewis Thomas's "tough membrane" was bacterial photosynthesis. From the beginning, it seems clear that geological evolution is the abiotic predecessor of life and its strategies but that ongoing geological change is in some part caused by the biota. This is a partnership that is prophetic of changes in ecosystems observed in historical times and now by ecologists.

Significantly, the pattern of succeeding ice ages alternating with warmer trends commenced nearly a billion years after bacterial photosynthesis was initiated. Early ice ages are, in fact, attributed to bacterial depletion of atmospheric carbon dioxide. Still later, certain host bacterial cells engulfed other bacteria within their membranes. At about the same time, the atmosphere was being oxygenated and converted from a reducing to an oxidating chemical mix—with devastating effects on the very life that had brought about this profound alteration. In the course of the oxygen revolution nearly 2 billion years ago, some bacterial species died, while others survived in dark anaerobic places. Still others entered into intimate symbiotic relationships whereby one was protected by and provided nourishment for the other.

It is believed that the mitrochondria in which cellular respiration occurs are the no-longer-independent descendants of the engulfed bacteria. In another 100 million years, protoanimals were evolving, and in another similar interval, blue-green algae followed. Geologically, oxygen in the upper atmosphere was forming a layer of ozone. Protoplants evolved in yet another 100 million years, and seaweeds followed in 500 million years. Here were two more innovations: multicellular structure and sexual reproduction, the latter allowing for greater diversity of organisms within a species than is possible with asexual reproduction alone. A series of ice ages followed. Some 670 million years ago, jellyfish and worms evolved. Their innovation is the hollow body formation that is the mark of familiar plants and animals today. The brain, a concentration of nerve cells in one end of every organism possessing it, evolved some 620 million years ago, and vertebrates followed in another 100 million years or so.

More ice ages followed around 440 million years ago, and somewhere along the line, life moved out of the water onto the land. Insects became prevalent 395 million years ago. Something more than 20 million years later, immense tsunamis struck, apparently attributable to a cosmic object hitting the ocean.

In the same interval, amphibians and trees appeared, and reptiles began to move about the land masses of the time. Reptiles held sway until the next period of ice ages some 290 million years ago, followed by another global catastrophe, this time attributed by some to the assembly of a supercontinent or to a collision with a comet. Nearly all species of marine animals were annihilated, and all the large terrestrial mammal-like reptiles were wiped out.

A scant 50 million years later, dinosaurs were taking their turn at dominance among animals, and the first flowering plants appeared. At 170 million years ago, the earth went through a greenhouse phase. This period represents a global boom in petroleum formation, the result of deaths among the immense biomass.

Birds appeared 150 million years ago. Shortly thereafter—by the geological scale—there was a major fall in sea level associated with another round of mass extinctions. But during this interval, the flowering plants commenced to support new animals, the mammals. Less than 100 million years ago, more than half the known global oil reserves were created as the continents flooded and submerged the vast growth of plant life.

Nearly 70 million years ago, primates were present, according to fossil findings. Yet another catastrophe occurred, during which the dinosaurs became extinct. It is not known whether a comet or giant asteroid struck Earth or whether the agent "killed quickly or slowly, by shock wave, poison, heat, cold, the obscuration of the sun, or prolonged ecological disturbance." Whatever the agent, it "dusted the entire planet with exotic chemicals. For about ten thousand years afterward the oceans were dead."

Another cosmic impact followed as the plant entered another cold spell, and the distinct pattern of summer and winter emerged. Twenty-nine million years ago sea levels fell, and on the emerging land, grass became prevalent—"the edible carpet." Ape and monkeys split into two distinct evolutionary lines, and antelopes appeared, the ancestors of domestic cattle and sheep.

Fifteen million years ago yet another cosmic object hit what is now Europe. At this time, Antarctica entered into a deep freeze that has persisted into the contemporary time. Intensive volcanic activity marked this interval. From 10 million years ago to nearly 7 million years ago, glaciers spread into Alaska and the southern portions of the North American continent. Following thousands of years of drying and flooding from spills from the Atlantic Ocean, the Mediterranean Sea normalized.

Five million years ago the common ancestor of contemporary gorillas, chimpanzees, and humans arose and life came up with another, quite different response to... geographic and climatic changes. In one primate line, an evolutionary adaptation to changing circumstances became an adaptation to every circumstance. The eventual outcome was an animal so unspecialized in its habits that it could live in any setting whatsoever.

ORIGINS AND HISTORY OF HOMO SAPIENS SAPIENS

Some 3.25 million years ago the current pattern of ice ages commenced in cycles of 90,000 years. In the course of this geological cycle, *Homo habilis*, the maker of stone tools, arrived. This species took up scavenging some 2 million years ago, and another species, *Homo erectus,* took up hunting in another 100,000 years. Fire making followed in 500,000 years, and *Homo sapiens* arrived in Europe 600,000 years ago. In another 50,000 years, the thirtieth of the current ice ages descended with warm interludes commencing 500,000 years ago.

In another 300,000 years, *Homo sapiens* domesticated an animal recognizable as modern cattle. A false ice age occurred in the relatively brief span of no more than a century, and shortly thereafter, the predecessors to the modern human race arose. After another false ice age, *Homo sapiens* sapiens was evolving in Africa. Geological upheaval—a major volcanic explosion, climatic oscillations, and the thirty-fifth modern ice age—intervened.

Then, some 60,000 years ago, humans were practicing herbal medicine, and 40,000 years ago, modern humans, the users of language, appeared. Soon thereafter humans began devising calendars, and the Neanderthal became extinct.

These new members of the genus *Homo* formed communities, tamed dogs, and survived the end of the ice age. Along the way, they began to practice decidedly modern techniques including irrigation, copper smelting, and the construction of stone buildings. Thus, humanity discovered and made use of what we call chemistry today. They also tamed horses and instituted taxes, the latter at least associated with the beginnings of class distinctions. By 5,500 years ago, two diverse inventions furthered the cause of human populations: the wheel and writing.

In historical times, humans have suffered the bubonic plague and the Black Death, but they invented printing. The astronomical telescope was created, and science was formalized during the seventeenth century. Oxygen was discovered as recently as 1771, and the germ theory of disease was presented in 1863.

The beginning of the fossil-energy revolution is marked at 1825. In 1885, the automobile became available to a few intrepid motoring pioneers. Antibiotics were discovered in 1940; gene structure, in 1953; and space craft were being built by 1957.

The green revolution in agriculture is placed in 1961, and gene splicing at the molecular level began in 1973. In 1977, *Homo sapiens sapiens* deliberately eliminated another species from the biosphere: The smallpox virus was eradicated.

MILESTONES IN THE HISTORY OF THE BIOSPHERE

Every person, whether environmental professional or simply having a personal interest in one or more levels of ecology, will isolate a few events that can be distinguished as milestones in planetary history. From the ecological perspective, every step in the global strategy is crucial to life, if not actually mediated by some portion of the biomass, and each step can represent a crisis in homeostasis.

Particularly critical events must start with the Big Bang, the ultimate source of all energy and material in the universe. From that ultimate source, the next critical step is the formation of the lithosphere of earth in combination with the cooling of the planetary surface to the range in which H_2O exists as a liquid.

The next critical steps were the origin of life and the incorporation of certain bacterial cells into others with the formation of membranes, separating life from nonlife. The oxygen revolution was perhaps the single most important ecological crisis.

The next critical step was the evolution of an organism that was to be the ancestor of primates, represented today by gorillas, chimpanzees, and humans. In terms of geological time and past events, rapid manifestation of crises has attended the development of the human species. Language, communities, and culture are a combination unique to this particular primate.

It is human culture that evolves, and rapidly, rather than the organism itself. Most recently, humanity has undergone or perpetrated a particularly critical step—the formalization of science out of a mixed background of magic, religion, and industry. The great crises for humans, then, are the advent of hunting and gathering, settling in communities as tillers of the soil, entering in rapid succession the industrial and scientific revolutions, and many would add, the information revolution associated with computer technologies.

Many would argue that the absolutely essential response to ecological crises, today at least allegedly anthropogenic in origin, is a restoration of humanity's connection to the biosphere. Nowhere is that restoration more significant than it is in terms of climate change.

Chapter 7

Greenhouse Effect

In order to examine contemporary issues associated with the so-called greenhouse effect in terms of anthropogenic causes, it is essential to place the effect in geological and climatological perspective. To do so requires a return to the matter of energy flow through ecosystems to explore the balance between entering and departing radiation, including heat.

The temperature range of the atmosphere is critical to, and maintained in large part by, the living component of the biosphere. Yet, as has been discussed at some length, the biosphere is itself evolving. Moreover, it is proceeding through cycles discernible in units measured by seasons, years, centuries, and the vast geological intervals. The latter include the "recent" planetary pattern of ice ages and intervals of warming.

The greenhouse effect of contemporary usage is asserted to be an anthropogenic perturbation that acts by interrupting the dissipation of energy from ongoing ecological processes and by the rapid utilization and resultant dissipation of the energy stored over geological time in the form of "fossil fuels."

That contemporary greenhouse effect is subject to substantial scientific, technological, economic, political, and legal scrutiny. But the issues and questions are not always sorted into those distinct categories. The questions before us can be stated in a kind of hierarchy from the large and impersonal to the highly personal: Is the perceived greenhouse effect a phenomenon of biosphere evolution and strategy, of lesser climatic cycles, or of very recent anthropogenic origin?

Or is it some complex amalgam of all those phenomena? Can direct and indirect effects of some potentially devastating, albeit transitory, trend of global warming be accurately predicted in terms of human activities, let alone in terms of ecosystems and the biosphere at large?

If such a trend is at least partially attributable to human activities, can those sources of the perturbation be isolated and identified generally and specifically? Then, the question, which goes beyond science, is whether we can prevent anthropogenic global warming or even reverse the perceived trend. Even further removed from science, although inextricably connected, is whether we *should* take action to halt or reverse the contemporary trend of global warming. What if, for one example with some precedent, we are entering another mini-ice age?

Have we the knowledge and the wisdom to determine whether prevention or reversal is either necessary or desirable, assuming either is possible? Can we make intelligent, informed choices among alternative responses? Assuming affirmative responses to the foregoing series of queries, have we the means to make the requisite choices through existing institutions of science, technology, politics, economics, and law?

Just as the questions range, so do the putative answers. Political, economic, and social responses make sense only when tailored in the first instance to the realities of the phenomenon. An imminent climatic change at a geological scale calls for very different adaptations in human populations than does a disproportionate meteorological alteration attributable to certain specific human activities.

Only when the choices have been so tailored does it make sense to apply them in the identified institutions of human endeavor.

ON THE GEOLOGICAL AND GLOBAL SCALES

Weather prediction has never been an easy endeavor. To attempt prediction, it is necessary to understand the atmosphere, itself an incredibly complex system of air and moisture in motion. Prediction can be categorized as a problem in fluid dynamics, probably composed of no less than hundreds of equations. The new science of chaos theory goes so far as to suggest that no matter how facile one becomes at solving the array of equations, we will find weather inherently unpredictable more than two to three weeks in advance.

If we cannot predict weather, how can we hope to predict global warming *and* determine its source? Have *any* of the pertinent questions been answered? According to James Hansen of the Goddard Institute for Space Research, two or three essential questions have been answered.

It is his position that no one can dispute that global increases in temperature have been statistically significant since 1965. That the warming is consistent with models predicting greenhouse warming is similarly undisputed. What is in dispute is the matter of cause and effect. In sum, any number of mechanisms could be behind the changes individually or in combination.

It makes sense to commence any responsible inquiry with the solar system and relevant earth sciences. These sciences reveal that incoming energy is radiation of short wavelengths, mostly in the range visible to the human eye. Atmospheric gases absorb this radiation only weakly. Outgoing radiation is mostly in the infrared, which is readily absorbed—and hence retained—by the atmosphere. Theoretically, without the atmosphere, global temperature would drop some 35°C.

While we still do not comprehend how solar variability influences either weather or climate, we do attribute geological patterns in terms of planetary orientations to the sun. At least in part, these patterns are related to variations in the tilt of the planet, the direction of its axis of rotation, and the shape of its orbit around the sun.

We also know that global temperature was 3° to 4°C warmer 5 to 15 million years ago and that over the millennia atmospheric carbon dioxide has been determined by the equilibrium of this gas between the atmosphere and the oceans. Significantly, atmospheric concentrations have been inferred to be 200 ppm during glacial intervals and 270 ppm during interglacial.

This difference can explain about half the difference of 10°C. Also significant is the understanding that the planetary norm is glaciations of some 100,000 years with interglacial separations, such as that in which we now live, of only a few thousand years. It appears that climate changes on this scale can be abrupt and that mini-warming or-cooling periods have occurred. "Little ice ages" are a phenomenon of recent history.

During past interglacial intervals, temperatures in North America were 2° to 3°C higher than they are now, and tropical species inhabited what are now temperate areas. At least some of these warm intervals are attributed to phenomena other than carbon dioxide concentration. It is worth noting in passing that seasonability has also been both stronger and weaker than is experienced at present.

Research and debate are associated with these phenomena, but no explanatory mechanism has been devised. Not only are cause and effect subject to controversy, but timing and duration of global climatic phenomena are also revised from time to time as new information or new interpretations come forward. There is even serious attention to the fact that every 180 years eight of the planets in the solar system are on the side of the sun opposite the Earth. Arguably, the result of that particular alignment could be earthquakes and extreme weather on Earth.

At the outset, it is worth noting that the study of anthropogenic climate change is only about fifty years old. Nevertheless, it is held probable that atmospheric concentrations of carbon dioxide were at 270 ppm in 1850 and began to rise with industrial burning of fossil fuels. Prior to that time, the concentration is believed to have been constant. From 1950 to the mid-1980s, the concentration increased from 310 ppm to 340 ppm.

In a comparable interval of 100 to 200 years, natural variation in temperature has been 0.5° to 1.0°C, but it is worth noting here that trends have been measured on a global scale only since 1880. Since then, decreases in temperature occurred between 1880 and 1930; the interval from 1930 to 1960 was marked by relatively stable temperatures; from 1960 into the

1970s, temperatures showed a small increase; and increases occurred from the 1970s into the 1980s. Adaptations of the biota to the climate of historical times have been possible because of relative stability. Thus far at least, little alteration in life spans from trees to humans have been noted. At least until now, plant and animal migrations have not been blocked by development, and human migrations involved relatively few individuals subjected to fewer political barriers.

Prognostication

Against this background, carbon dioxide concentration is now at 350 ppm and increasing by some 0.4% each year. On an annual basis, some 5 to 6 billion tons of carbon are released from the burning of fossil fuel into an atmosphere containing some 700 billion tons of carbon dioxide. Some carbon, of course, is stored in the biota and lithosphere, but simultaneously with the burning of fossil fuels, an unknown quantity of deforestation is occurring. This loss is also a loss of biotic carbon dioxide assimilation, allowing further increases in atmospheric concentration.

On a smaller scale, respiration is increasing as temperatures increase. Calculations indicate that the rate of respiration is increasing faster than that of photosynthesis. The latter is affected primarily by the availability of light, water, and nutrients as well as by the presence of chlorophyll, whereas the former, while dependent on available water, is particularly sensitive to temperature.

Those who engage in prognosis assert that for each doubling of carbon dioxide concentration, a temperature increase of 3.5° to 4.5°C follows. For every global increase of 1°C, temperature zones shift by 100 miles. By the year 2030, a 2°C increase is then predicted to move these zones 200 miles.

Migrations to suitable habitats could occur except where blocked by natural phenomena such as oceans and mountains and by social or political features such as agriculture and urbanization. Among human populations, about one half of all individuals inhabit coastal areas, which are expected to be inundated.

The Uncertainties

The considerable uncertainties associated with the contemporary greenhouse effect are attributable to incomplete scientific understanding of the array of relevant phenomena. Over the last several hundred thousand years, there has been nothing analogous to today's greenhouse gases. In order to make predictions, inferences are required. One approach takes the form of climate models, by which present events are calculated for comparison with actual experience. Where reconstructions of the past are available, those events are computed for comparison to present experience.

Another approach is to use the planets Mars and Venus for comparison, the one with less and the other with more capacity for trapping heat than Earth.

Whatever the quality of such predictions, it is essential to integrate them with other geophysical phenomena. In one case, the influence of sunspot activity, the effect on climate is simply unknown in terms of presence, absence, or extent. The phenomena dubbed El Niño and La Niña are relatively new to climatology and thus something of mysteries in themselves. Then there is the matter of volcanic activity and its effects on global climate, including temperature.

In the end, how can decisions be made on behalf of human populations, let alone the biosphere, when short-term global temperature changes may in fact be headed up or down naturally, and may as well be tempered as exacerbated by anthropogenic events? This is the uncertainty facing those responsible for policy, especially policy affecting ecosystems, and for the environmental professionals who serve the policymakers.

The first step is a familiar one—the unpopular but essential need for expanded research efforts. Popular or not, it is essential to know more of large-scale oceanic circulation, of phenomena as disparate as processes regulating soil moisture and those responsible for cloud formation, of the influences of biogeochemical cycles on atmospheric composition, and of processes regulating sea ice. When we have a better grasp of the underlying scientific realities, we can turn that knowledge to such crucial matters as the choice between adaptation and prevention—or even the much-maligned "do nothing" or "wait and see" responses.

It is often said that greater harm will follow if we mistakenly take one of the latter approaches in the face of perceived environmental threats than if we are mistaken in an active response such as the former approaches. But here is a situation where doing something at tremendous expense to institutions and individuals itself could go beyond any such harm to exacerbate the effects of natural trends that were never imagined, let alone predicted.

That prospect should cause us to be less inclined to dismiss scientific uncertainty in the rush to do what is momentarily deemed the right thing. That is a telling incentive for societal scientific literacy, especially in ecology and the environmental sciences—at least to the point of appreciating the implications of uncertainty.

THE BIG BANG AND BLACK HOLES IN SPACE

At the beginning of this chapter, our ability to predict our ecological future from the foundation of relevant sciences was called into question. Among any number of practical problems of inadequate or inaccurate prognostication is that associated with global warming. Proponents of the Gaia Hypothesis might assert that it does not matter what we do: Whatever it is will be the mechanism intended by Gaia for survival of the planet. We recognized the evolution of the biosphere to be directional, part of a global strategy over which human institutions may actually have no control for better or for worse. In terms of the strategy of our species, each of us must decide how we will respond to that knowledge. Global marketplaces and governments are, after all, populations within populations.

Each is made up of individuals. We are only just beginning to learn how to adjust our Behaviour to maintain our microcosms on behalf of our ecosystems. We are only just beginning to glimpse possible biospheric perturbations' brought on by our numbers, distribution, and consumptive habits. We would do well to take immense care in predicting where we and our fellow members of the biosphere are going, especially when we would take upon ourselves redirection of the perceived strategy. We should act with the humility demanded by the immensity of what time's arrow has wrought and holds for the future. The relevant times are infinite, at least from the limited perspective of a human lifetime. Even the duration of human presence on earth is humbling in the face of cosmology.

Whatever we do, however long our species endures, the ultimate fate of the *planet* is not in our hands. Earth's fate is linked to that of the sun. To contemplate the cosmos is to reinforce the lesson of humility.

It is possible, according to cosmologists, that every black hole—a star so immense it collapsed into infinite density from which neither gravity nor light (and hence spacetime) can emerge—is another big bang. Thus, there may be parallel universes with parallel solar systems with their respective planets Earth. One can speculate, along the lines of decades of science fiction, that some of the species corresponding to humans have done better than we in discovering and applying ecology to their own biospheres.

On the other hand, they may have done far worse—or had no lasting impact at all. Should a black hole someday swallow up our galactic neighborhood, what we have done to or for this planet is hardly going to be particularly significant. It is time we developed a better sense of our place in the natural world. We leave the realm of the natural world for those of philosophy and theology when we would usurp the omniscience and omnipotence prerequisite to running biospheric strategy.

Chapter 8

Evolution of Ecosystems

When turning from biospheric strategy to the counterparts at the ecosystem level, additional perspective on the ecology of human populations is provided. We have encountered the argument that any fragility in life as a planetary force is illusory. There is no need for resort to Gaia in arguing for the toughness and longevity of life. What may be far more delicate is our own species.

PLANETARY ECOLOGICAL DEVELOPMENT

Returning for the moment to particle physics, the ecologist is advised that slightly stronger nuclear forces would render hydrogen rare. With a reduction in the amounts of the simplest element, long-lived stars like our sun could not exist, and there would be no water, if they did.

Slightly weaker nuclear forces would preclude nuclear ignition of hydrogen. Without ignition, no heavy metals would emerge, and carbon-based life could not exist. Without the exclusion of one electron by another, there would be no chemistry.

Similar accidents or coincidences abound throughout the sciences, and they all hold implications for ecology. Among these coincidences are those that allow water to exist as a liquid, carbon atoms to form complex organics, and hydrogen atoms to form readily breakable bonds between molecules. In almost every case, no more than a slight increase or decrease in a single parameter would render life not different from that which we acknowledge as such but *impossible*.

Once life emerged, the strategy of the biosphere could not

fade out of being. Instead, life itself imposes the strategies requisite to its own continuation "against the mischief of the world." In the consolidation of apparent accidents and coincidence as a discernible strategy of the biosphere, the ecologist might well return to the boundary between life and nonlife, the cell. Not only is the cell the structural and functional unit of all life; it stands representative of those levels of organization that are biological systems.

Metabolism, occurring within the cell, requires the same kind of delicately balanced environment that consistently distinguishes the biotic from the abiotic. "Cells are the life-support chambers that contain this special environment. A living cell keeps its chemical composition steady within narrow limits, a condition known as homeostasis." It is the cellular membrane that separates the living cell from its environment. The evolution of the membrane may then be the final step in the emergence of a biota from a planet of geological processes described exclusively through physics and chemistry.

This ultimate transition in the planetary continuum stretching all the way back to the Big Bang and indefinitely into the future is obviously a crucial step in planetary strategy. Earth may truly be unique in the universe.

The unique agent of planetary strategy is now life, and all of life is itself a continuum linked by the molecules of inheritance. Astonishingly, our DNA is linked in an unbroken sequence from the same molecules in the earliest cells that formed at the edges of the first warm, shallow oceans.

Our bodies, like those of all life, preserve the environment of an earlier earth. We coexist with present-day microbes and harbor remnants of others, symbiotically subsumed within our cells.

While life is maintaining its identity and its separation from the nonliving, it is also maintaining the biosphere. If this is indeed a strategy, it is particularly obvious in the atmosphere. First, in the distant past, life brought on a global pollution crisis, "the oxygen holocaust." Through photosynthesis, atmospheric carbon dioxide was depleted at a juncture when molecular hydrogen was escaping and hydrogen sulfide becoming insufficient.

Life accommodated to the changes it had brought on even as it was stabilizing them. Water, the abundant dihydrogen oxide, has replaced hydrogen sulfide in photosynthesis, with a release of oxygen instead of sulfur. But oxygen is toxic because of its high reactivity with organic matter! Nevertheless, life adjusted so successfully that atmospheric molecular oxygen went from a concentration of well under 1% to 21%. The cyanobacteria initiated the controlled combustion—oxidation-which is aerobic respiration.

Strategy or mere coincidence, molecular oxygen in the atmosphere holds today at 21%. Were it lower, aerobic organisms would suffocate. A few percent higher and living organisms would undergo spontaneous combustion. Strategy or coincidence, it is oxygen that "filters out the very bands of ultraviolet most devastating for nucleic acids and proteins, while allowing the penetration of the visible light needed for photosynthesis."

Strategy or coincidence, the ozone layer now protects living entities from mutation, which is perceived today as a hazard to life. Before that protective layer was generated, living entities were undergoing frequent mutations leading to evolution through speciation, perceived today as beneficial because the processes have resulted in the familiar biota of our own time.

There is no reason to believe planetary strategy has ceased. The process is, however, too slow for contemporaneous measurement by individual human populations. A comparable strategy can be perceived among ecosystems. Such strategies are not merely significant in terms of immediate environmental conditions. They are a continuation of and a mediator of biospheric strategy.

ECOSYSTEM STRATEGY

Just as the planet and its biosphere have manifested a direction in their lengthy and partially mutual histories, each ecosystem manifests a strategy. If there were no such direction with discernible steps theoretically leading to the so-called climatic climax community, there would be no biomes by which terrestrial ecosystems can be classified. Furthermore, such a strategy informs of diversions, with special reference to those caused by anthropogenic intrusions into ecosystems.

A major long-term strategy behind ecosystem development is coevolution involving two or more organisms. Distinct communities arise because certain plants and their herbivores have evolved in mutual relationships, as have parasites and hosts and the microbial symbionts of a wide variety of animals—including humans as well as their domestic species. In fact, today's symbionts can be viewed as particularly successful parasites. Whenever two or more organisms act to their mutual benefit, each is more likely to have a multigeneration future than is possible when one is lethal to another.

Aquatic

Life is believed to have originated in the oceans. Recent evidence suggests that the geological processes from which life arose there still exist. Oceanographers have observed the mingling of magma, steam, and gases at underwater seams between continental plates. Where the cold, dark ocean floor is mostly barren, at these seams communities based on sulfur bacteria thrive.

In fact, succession continually occurs in aquatic environments but is far less obvious to the casual observer than in terrestrial ecosystems. Marine ecosystems are apparently well established, yet fluid, with notable succession occurring primarily where land meets sea or fresh water meets saline. Because of geological alterations, such areas tend to be in flux with predictable ecological accommodations adapting to altering habitats and niches.

Examples of such alterations are the slow, ponderous movement of continents, the "traveling" of barrier beaches, volcanic upwellings, and alterations brought about when severe weather makes landfall.

In freshwater ecosystems, succession is mediated by two forces in tension with each other. Geological features provide direction, while the biotic tend to slow the resultant processes or to stabilize the ecosystem. Geological forces cause erosion and the filling of lakes, altering streambeds and contiguous terrestrial habitats and carrying a lake through the process of eutrophication, resulting ultimately in a terrestrial habitat in which invasion and succession will proceed.

In a sense, succession in the waters of the planet is heavily influenced by geology and hydrogeology and forms a part of terrestrial strategy as well. This is especially the case over lengthy intervals. Aquatic communities, then, seem to relate far more to place and a human time frame than to geological time.

Terrestrial

A succession of organisms and of influences between organisms and the abiotic environment occurs in both aquatic and terrestrial systems, but the terrestrial systems are those that tend to define a planetary region in ecological terms. Each terrestrial ecosystem starts with an array of abiotic conditions. First and foremost are those of the climate, dependent on latitude and altitude.

Organisms invade and succeed one another, as temperature and moisture regimes allow, but success and succession depend on the substrate—the chemical composition and physical state of the land—although it is itself modified by the resident biota over time.

Because terrestrial succession takes the form of an orderly process of community development, it is reasonably directional and, therefore, predictable. Succession results as invading organisms establish themselves while modifying the physical environment, and those organisms and their modifications make way for other organisms. In a series of steps known collectively as the sere, community composition alters along predictable lines. The abiotic environment determines the pattern, establishes the rate of change, and may set limits on how far the sere can go.

It is a changing community, however, that controls the actual process. The sere is expected to culminate in a stabilized ecosystem, which will remain in place as long as the climate does not alter and no perturbation seriously alters the homeostasis.

The relatively transitory communities are recognized as seral stages, also known as developmental or pioneer stages. The terminal stabilized community is known as the climax community. The relatively short-term strategy of an ecosystem is comparable to that of the biosphere over the long term. Obviously, the two strategies are inextricably interconnected.

Criteria of succession have been identified. They are community energetics and structure, organismal life histories, nutrient cycling, selective pressure, and overall homeostasis. As succession progresses, the effects of production and respiration become balanced with the generation of greater biomass over time. A relatively large biomass comes to be supported for each unit of energy. Net production, however, is relatively low, and there are food webs rather than simple food chains. The detritus cycle becomes a primary component of the whole.

With the alterations in energy relationships, community structure is altered toward a large total component of organic rather than inorganic matter. Inorganic nutrients are found in the biomass. Both taxonomic and biochemical diversity increase in well-organized patterns of distribution and Behaviour—niches become intricately distributed. An increased variety of species accompanies a reduced dominance by one or a few species.

As the community so alters, individuals are subject to characteristic modifications. Narrow niche specialization is accompanied by the appearance of large organisms with lengthy and complex life histories. Nutrient cycling undergoes related alterations. Mineral (inorganic) cycles are more closed than open, and nutrient exchange within the biota slows. As these alterations occur, the detritus becomes increasingly important in the biogeochemical cycles.

Selective pressures are affected by feedback controls on

growth and enhanced quality in production, often expressed as diversity or stability. Overall, homeostasis does become relatively stable, that is, a steady state exists.

Internal symbiosis has developed, nutrients are conserved, and the system becomes resistant to perturbations. In short, the system is subject to low entropy and maintains a high level of information. The overall strategy is believed to be one of movement toward as large and diverse an organic structure as possible within the limits set by available energy input and prevailing physical conditions.

The climax community is the theoretical—and sometimes actual—final, stable community for each sere. This community is self-perpetuating and in equilibrium with the physical habitat of the ecosystem. The theoretical single climax community of a given global setting will be realized wherever local physical conditions are not so extreme as to modify the controlling effect of the regional climate.

In many locations, that theoretical community will not become established. Instead, there will be a pattern of edaphic climaxes, the perpetuation of one or more communities intermediate along the sere. Another pattern is that of the catastrophic or cyclic climax, often maintained under the influence of humans or their domestic animals.

Two kinds of succession are recognized: primary and secondary. Primary succession is the natural process manifested in an area not previously occupied by life. Secondary succession is that which proceeds upon the removal of an existing community, natural or anthropogenic.

IMPLICATIONS FOR ENVIRONMENTAL PROFESSIONALS

In some ecosystems, the sere pauses or halts indefinitely in one or more edaphic climaxes. A variety of conditions will preclude progress to the climatic climax while allowing the maintenance of a relatively stable one.

Climate itself may represent seemingly minor differences from one location to another. On one scale, north-and south-facing slopes on the same elevation will support noticeably different plant communities. On another scale, microclimates in soil temperature and moisture regimes will support correspondingly distinct communities of microorganisms.

Between the two extremes there will be unique plant communities. Similarly, soil and communities of larger organisms are affected by the intricate physicochernical relationships among soil chemicals and particles, including the chemistry of products; pH; and mobility of elements, ions, and more complex substances.

Edaphic climaxes are naturally occurring. A disclimax is the community resulting from a human intervention that prevents the sere from reaching the theoretical climatic climax. (A less accusatory description is anthropomorphic subclimax.) Excessive harvesting or overgrazing by domestic herds are well-recognized disclimaxes.

In addition to the foregoing, cyclic or catastrophic climax communities are recognized. The agent here may be a biological one, but the best known is fire. Fire, often caused by lightning, routinely clears forests of thick underbrush. Not only does this clearing maintain the community and promote the reproduction of certain populations, but it also acts as a preventative. Frequent brushfires prevent the devastation of crown fires.

Thus, the ecological aftermath of brushfires is not mere renewal or restoration, as is the case with a crown fire, but a component of ecosystem strategy. Although recognized by ecologists as such, forest fire has been a point of major contention among environmental professionals. The Yellowstone fires of the late 1980s brought the controversies to the political process and to the headlines.

Chapter 9

Abiotic Influences on the Biota

With regard to environmental constituents, the organisms in an ecosystem either are in a transient state, during which the constituents of the surrounding environment are rapidly changing in some particular way, or are in a steady state. While relatively constant, a steady state should not suggest stasis among or within organisms. The only static biological system is a dead one! The condition is better described as a kind of dynamic equilibrium whereby the inflows of energy and materials balance the outflows in terms of biological effects and their quantification.

As has been suggested throughout this text, every equilibrium is the net sum of many distinct but connected chemical reactions. First and foremost among them are photochemical reactions. All the rest are thermochemical reactions. Not surprisingly, photochemistry is driven and limited by light intensity. Thermochemistry, as the name implies, is driven and limited by temperature regimes.

Since each observation of the ecosystem is in effect a snapshot, the steady state as well as one recognized as transient may be ephemeral. Living organisms are notoriously resistant to being pinned down. Thus, the discussions to follow may be more a reflection of limitations on observation than of the richness of detail in ongoing natural processes.

Nevertheless, the phenomena associated with limiting factors, ranges of tolerances, and interactions among them provide essential detail in connecting organisms to the abiota. Moreover, the added detail informs comprehension of associations among organisms.

As a point of departure, terrestrial, marine, and freshwater ecosystems are primarily subject to characteristic regimes of interacting factors. For terrestrial ecosystems, the regime is light, temperature, and water, the latter in terms of both precipitation patterns and availability. The marine regime is light, temperature, and salinity, whereas that for freshwater ecosystems is light, temperature, and concentration of molecular oxygen. All ecosystems are regulated by the availability and distribution of mineral (inorganic) nutrients and their rates of movement through the biogeochemical cycles.

LIMITING FACTORS

Limiting—or regulating factors are best discerned when an ecosystem is more accurately described as being in steady state than in a transient one. Odum has proferred two "laws" to describe the effects on organisms of limiting factors. The first is Liebig's "Law" of the Minimum; the second, Shelford's "Law" of Tolerance. Both principles are stated as *quasi*-laws, because they do not rise to the level of the relatively fixed concept of a law in the usual scientific sense.

Rather, each constitutes no more than a vastly simplified expectation of a specific ecological relationship, which may be manifested at least briefly despite the daunting complexity of real interrelationships and despite the ephemeral nature of any such relationship. Regardless of the imposing caveats, the two laws remain worthy of attention because they do assist in understanding events in an ecosystem.

According to Liebig's Law of the Minimum, when an ecosystem is subject to steady-state conditions, the essential material, or other factor, available in the amount most closely approaching the critical minimum for the subject biological system will tend to be the factor limiting that biological system. Limiting factors are usually analyzed at the level of a species or population or of the individual organism, although the immediate and most direct effect may be occurring at some other level of organization.

As might be expected, that limiting factor is not often

among macronutrients. More subtle factors tend to be limiting. This law is said to be less applicable, not inapplicable, under transientstate conditions.

It is likely, however, that it is the observer's inability to detect and measure such fleeting limitations when conditions are rapidly altering. In all likelihood, one and then another and still others will be briefly limiting as environmental conditions change and organisms respond to the changes.

Indeed, comprehension of limiting factors generally is frustrated by the very dynamic interactions that constitute ecosystems and their components. For example, if the observer is monitoring modifications in the rate of utilization of a factor believed to be limiting, the effort can be frustrated by such matters as the high concentration or availability of another substance or factor. Such naturally occurring phenomena alter biological responses from the molecular level to the level manifested as Behaviour.

Moreover, the observer may not even be aware of the presence, let alone the influence, of another factor. From the perspective of the biological system, there is the matter of the tremendous ability to substitute, often at the level of molecular biology, where alternative metabolic pathways may be available, depending on chemical and physical circumstances.

TOLERANCES AND RANGES

Shelford's Law of Tolerance is probably the more accurate reflection of natural complexity. It holds that first the presence and then the success of an organism depend on the completeness of a complex of conditions. The absence or failure of an organism, then, is a function of qualitative or quantitative deficiency or excess with respect to any one of several factors approaching that organism's limit of tolerance for it.

More precisely, each organism—whether the individual or the species population—is subject to an ecological minimum, maximum, and optimum for any specific environmental factor or complex of factors. The range from minimum to maximum represents the limits of tolerance for the factor or complex. Significantly, if all known factors are apparently within their respective ranges for the subject organism and yet it fails, it is necessary to consider additional factors or a more complete array of interrelationships, including interactions with other organisms.

When faced with any such situation, it is essential to remember yet another caveat, a biological reality to be taken seriously: "Studies in the intact ecosystem must accompany experimental laboratory studies, which, of necessity, isolate individuals from their populations and communities." Put another way, it is essential for field biologists to consider the wisdom to be discovered in the laboratory and for biologists ordinarily bound to the laboratory to be aware of the ecological reality associated with the processes they are investigating.

There are significant corollaries to the Law of Tolerance. An organism may have a wide range of tolerance for one factor but a narrow range for another. Logic suggests that organisms with wide ranges of tolerance for many, if not all, factors will be the most widely distributed. Within an organism or species, when conditions are nonoptimum for one factor, limits of tolerance for others may be narrowed. Tolerance is most likely to be limited during periods of reproduction.

Evolution of narrow limits constitutes one form of specialization and reflects greater efficiency, but it does so at the expense of adaptability. These are the kinds of factors that contribute to increased diversity in the community as a whole.

Adaptation as such is ordinarily physiological, temporary, and readily reversible. A genetic alteration is more likely to be permanent and nonreversible. Such a genetic alteration is a step toward speciation and Darwinian evolution. A particularly interesting phenomenon is the socalled genetic fixation of local types within any taxon, but especially prevalent at the species level. Isolation of such a type from the parent stock can also result in speciation.

On the practical level, limits of tolerance provide an opening to the ecologist or to the environmental professional for studying and comprehending complex ecological systems. These conditions allow a determination of factors that are operationally significant to a biological system and of how factors can affect the individual, the population, or the community. The methods include observation, analysis, and experimentation.

Wisdom further dictates not only that the observer move back and forth between lab and field but also that each professional interact with specialists from other disciplines in the relevant sciences and technologies. Unfortunately, these are ideals that were not being met two decades ago:

In the training of biologists during the past 40 years, there has been an unfortunate cleavage between the laboratory and the field, with the result that one group tended to be trained entirely in laboratory philosophy... while another group tended to be trained quite as narrowly in field techniques. Modern ecology, of course, has become especially relevant to our times because it breaks through this artificial barrier and provides a meeting ground for the biochemist and physicist on the one hand and the range, forest, or crop manager on the other!

It is up to practicing environmental professionals to minimize any such cleavages, with special reference to the natural sciences in relationship to social and political sciences and practitioners of law. The days of balkanization among the environmental disciplines should have been left far behind. Never has modern ecology been so relevant to the times as it is now and in the foreseeable future. That the disciplinary barriers are artificial in terms of ecological entities is becoming increasingly apparent.

One impetus for better integration of other sciences and a variety of analytical approaches to ecology into the environmental sciences is the matter of interactions. No limiting factor exerts its effects in isolation. Tolerances and ranges influence each other, often altering the extent of limitation if not the identity of "the" factor limiting a biological process.

Interactions

Interactions of predominant and therefore obvious significance are those among temperature, moisture, concentrations of oxygen and carbon dioxide, and the availability of nutrients. Obvious or subtle, these influences may well be significant at any level from the subcellular to the biospheric.

Nevertheless, it is always wise to retain the notion that, ultimately, these dynamic events are actually biochemical in nature, determined and controlled by the genetic complement of the individual and that of the species-the gene pool.

Limiting or regulating factors in ecosystems can be placed in classes. In doing so, however, the interconnectedness of their respective influences must always be kept in mind. For example, light and temperature are important limiting factors affecting photosynthesis.

When light is operating as a limiting factor, the process is relatively insensitive to temperature. In light near a species' optimum, the rate of photosynthesis will increase as much as fivefold for every temperature increase of 10°C. In light near the maximum for the species, its range of tolerance for temperature extremes will be narrowed. In fact, each species will possess a characteristic optimum combination of light and temperature. With regard to light intensity, there are two steps of significance to the relationship between photosynthesis and its metabolic partner, respiration. First, there is the compensation point, the intensity at which the two processes are in balance.

The oxygen released in the course of photosynthesis is equivalent to the oxygen assimilated through respiration. Second, there is the saturation point, the intensity at which the photosynthetic rate no longer responds to increasing intensity. The biological system is "saturated" with light and cannot effectively utilize any more of that form of energy. Perhaps the most important factor is water or moisture. Water is the physical, chemical, and biological entity that is a physiological necessity for all protoplasm. Next to water, the most important factor for life is temperature.

The range of tolerance for life is astounding—-200°C to + 100°C, well below the freezing point to slightly above the boiling point of water. Most species, however, are individually far more restricted in their tolerance for temperature extremes. Generally, the upper limit is more critical than the lower, as might be inferred from the sensitivity to temperature of biological molecules such as proteins, especially enzymes. Paradoxically, organisms tend to function more efficiently as the upper limit is approached.

Thus, a stimulation of biological processes can occur, rapidly followed by morbidity and mortality as the upper limit of tolerance closely follows.

Populations of any given species are likely to be found in areas where there is a relatively narrow range of temperature at some characteristic point along the extensive range encompassing all species. That range is smaller in water than on land. A variety of adaptations accommodate the temperature range. Some species, especially among microorganisms, are able to go into a kind of suspended animation, or dormancy, allowing their survival for lengthy intervals at a temperature that would otherwise be lethal.

In summary, temperature is universally important to life and often a limiting factor. As a result, temperature regimes can be responsible for zonation and stratification, respectively, the horizontal and vertical distribution of organisms in the ecosystem. Nevertheless, it is easy to overemphasize the importance of temperature simply because it is among the easiest environmental factors to measure.

As might be expected, the most obvious complex of interactions among environmental factors is found in regimes of combined temperature and moisture. On the global scale, these regimes are represented by climate. But regimes of temperature and moisture also occur on decreasing scales, down to microclimates affecting microcosms measured in inches or millimeters. It is worth noting that temperature becomes more severely limiting—that is, the range of tolerance will be reduced—at extreme moisture conditions.

Other interactions occur among atmospheric gases in terms of their occurrence and availability in soil as well as in water. While atmospheric water Vapour undergoes large variations over space and time, the portion of the atmosphere that is identified with the biosphere or ecosphere tends to be stable or remarkably homeostatic. The relationship between carbon dioxide and oxygen tends to be limiting to many of the higher plants. Within known limits, moderate increases in carbon dioxide tend to increase photosynthesis, as do decreases in oxygen. In fact, oxygen inhibition of photosynthesis can occur, a negative feedback mechanism believed to be effected by the reversal of a chemical reaction between oxygen and a metabolic intermediate in the overall photosynthetic process.

Oxygen concentrations can be severely limiting in both soil and water ecosystems, although the chemistry of oxygen in combination with that of carbon dioxide is quite distinctive in water. Here, pH is a crucial factor. Biological systems are subject to specific ranges of tolerance for pH, which not only exerts a direct effect on the processes of life but also alters the chemistry of important nutrients and the toxic properties of chemical substances, including those that are macro-or micronutrients. Immediately, yet another pattern of interaction comes into play in aquatic ecosystems. Temperature is a determinant of concentrations of a gas in water. Chemistry, specifically the nature and quantities of dissolved and suspended substances, is another.

Ints and water pressure determine the distribution and local concentrations of gases and of nutrients as well. Populations adapted to local conditions, especially at boundaries separating ecosystems, can take on traits unique to them. Such populations are known as ecotypes. As Odum has noted, "The possibility of genetic fixation in local strains has often been overlooked in applied ecology; restocking or transplanting of plants and animals may fail because individuals from remote regions were used instead of locally adapted stock."

Such patterns will have noticeable effects on communities, which will respond with distinct morphological and physiological adaptations. In terrestrial ecosystems, wind plays a role comparable to water currents. Moreover, wind can cause a loss of water through increased transpiration, perhaps causing water or a nutrient carried in water to become limiting to a plant. Morphological, physiological, and behavioral adaptations to both extremes in the availability of water have evolved in terrestrial plants and animals.

Table. Elements Essential to Life

Essential	*Photosynthesis*	*Other Metabolic Functions*
Iron	Manganese	Manganese
Manganese	Iron	Boron
Copper	Chlorine	Cobalt
Zinc	Zinc	Copper
Boron	Vanadium	silicon
Molybdenum		
Nitrogen Metabolism Molybdenum Boron Cobalt Iron		
Chlorine		
Vanadium		
Cobalt		

Similarly, the biogenic salts, most notably those of nitrogen and phosphorus, represent chemical regimes in which organisms must function. Salts are the chemical products of the neutralization of acids and bases. Both dissolve readily in water and dissociate into positive and negative ions, which recombine to form salts, products that are substantially less soluble than the reactants.

The negative ion of the acid combines with the positive ion of the base. For example, hydrochloric acid (HCl) and sodium hydroxide (NaOH), a base, both dissociate in water to release the ions H_3O^+, Cl^-, Na^+, and OH^-. From these ions, sodium chloride (table salt, NaCl) is produced. (The hydronium and hydroxide ions also combine to produce water.) The nitrogen salts are crucial to life, and salts of phosphorus are among the most frequently encountered limiting factors. The macronutrients also include salts of potassium, calcium, sulfur, and magnesium. Of course, it is the micronutrients, components of enzymes and coenzymes—notably the trace elements—that may be limiting and for which ranges of tolerance may be small. This is an explanation of why a heavy metal is both essential to life and extremely toxic.

The metal molybdenum, for example, can be limiting to the whole ecosystem. An added concentration of 100 parts per billion (ppb) in a mountain lake can increase photosynthesis. But at the same time, the resultant concentration is high enough to be inhibitory to the growth of phytoplankton. The *net* effect here would be crucial:

Would increased photosynthesis or decreased growth prevail, or would the two in effect cancel each other out with no *net* change? In this context, it is essential to be aware of the influences of such unpredictable factors as storms and hurricanes. Not only will these meteorological factors affect the distribution of material; they can affect organisms directly. Protective Behaviour patterns have evolved. In addition to all the foregoing, there are factors such as currents and pressures exerting their influences on organisms.

Finally, soil itself and fire represent additional interacting factors affecting organisms. For terrestrial organisms and the benthic communities of aquatic ecosystems, soil (or sediment or other substrate) represents a crucial combination of interacting abiotic factors. In addition to inorganic and organic content, such factors include the size of particles composing the soil and the resultant texture and porosity. Fire is an important limiting factor.

IMPLICATIONS FOR ENVIRONMENTAL PROFESSIONALS

The Practical Side of Limiting Factors

The matters of limiting factors and, especially, tolerance ranges have major implications for ecologists and for environmental professionals alike. Well-established limiting factors can be put to practical use.

Not surprisingly, regimes of combined temperature and moisture are of particular practical import. Climographs, polygons, take form when monthly average moisture is plotted against monthly average temperature and the twelve points connected. When plotted for specific regions and then combined with a corresponding climograph representing tolerances for a population, certain questions can be answered (at least prima facie). Among them, What climax community is expected? Where can a species be successfully introduced? In what years will a pest be particularly active?

Climographs for the coastal and northern regions of some states are strikingly different. They never overlap. One region is subtropical with a pronounced wet and dry season, while the other's seasonal temperature differences are more pronounced than seasonal patterns of rainfall.

The climax community for the former region is broad-leaved evergreen forest of live oak and Spanish moss; that in the latter is temperate deciduous forest. Obviously, if one or the other community is not found, some factor, natural or anthropogenic, could represent a perturbation, once other natural factors have been taken into account.

In the case of the deliberate introduction of an alien species, the climographs for the two other states largely overlap that of the species' natural habitat. One state, however, is hotter, with greater rainfall from spring to summer. The other is cooler, with comparable rainfall in the winter. The species was successfully introduced only in the second state, suggesting two possible explanations:

Either the species is more tolerant for cool temperatures than for a hot, wet season, or the latter conditions were more extreme than the former. In the case of a pest organism, both optimal and favorable climographs were available for comparison to climographers drawn for two different years in a region where oranges are an important crop. In one year, regional conditions were always drier than the optimum, sometimes with cooler temperatures.

In one month, it was drier than favorable for the population. In the other year, both conditions were usually within the optimum range. For part of the year, temperatures were cooler but remained within favorable limits. In this second year, as would be predicted by comparing climographs for the pest and for the region, damage to the orange crop was much more severe than in the first.

The Interdisciplinary Nature of Ecology

One of the strengths of today's science of ecology is its demolition of artificial barriers among scientific disciplines. As ecological interactions are explored, the ecologist is hopelessly isolated if he or she is not engaged in an occasional meeting with a physicist or a biochemist. The environmental professional goes even further and may actually be responsible for the management of a crop, forest, or rangeland.

That professional must not only be steeped in ecology and related natural sciences; management implies the addition of at least economics, politics, and law, if not sociology and even psychology. And superimposed over the whole are the ethics of each profession as well as personal and even national environmental ethics. It is essential that each piece in this management puzzle be afforded due respect.

HOMEOSTASIS AND CYBERNETICS

By now it should be abundantly clear that organisms are no "slaves" to the environment in which they find themselves. Organisms are constantly adapting themselves *and* modifying their environments so that limiting effects are at least accommodated if not alleviated or eliminated altogether. Organisms and other biological systems are constantly seeking the stability known as homeostasis. They do so by internal cybernetic processes.

From this situation, it is easy to fall into one of two logical traps: teleology and anthropomorphism. Teleology is the fallacy of attributing sentient intent where it does not exist. Anthropomorphism, equally fallacious, goes further to impose human intent or sensibilities on nonhuman entities either abiotic or biotic. To the conservative scientist, Gaia crosses at least the teleological line. But even in doing so, the concept still provides valuable insight. At the cellular level, decisions as to the edibility or otherwise of things encountered, and as to whether the environment is favourable or hazardous, are vital to survival. They are, however, automatic processes and do not involve conscious thought. Much of the routine operation of homeostasis, whether it be for the cell, the animal, or for the entire biosphere, takes place automatically, and yet it must be recognized that some form of intelligence is required even within an automatic process, to interpret correctly information received about the environment.

Each level of biological organization engages in processes whereby that stability known as homeostasis is maintained. Even a minute change in the environment will trigger mechanisms that act to restore that stability or to establish a new homeostasis as conditions require. These mechanisms are the subject of the science of cybernetics.

Every biological system is capable of alteration through cybernetic mechanisms to meet environmental modification. But here, too, there are limits and ranges of tolerance. When those are exceeded, the system has been perturbed. Here, there are also ranges: Some perturbations can be accommodated; others will disrupt a biological system to the point of death.

With particular reference to energy in the ecosystem, there is a distinct pattern of considerable relevance here. The tendency in such a natural, enclosed system, through which energy is flowing, is for the community members to undergo shifting alterations and mutual adjustments leading to a degree of stability. These directional shifts, alterations, and adjustments actually are occurring in each affected individual organism through internal, self-regulating mechanisms. Each mechanism—physical, biochemical, physiological, or behavioral—tends to return the biological system to the constancy of homeostasis. Energy transfers in the unperturbed ecosystem in homeostasis tend to flow in one direction at a steady rate that is characteristic of the specific ecosystem.

In combination, these relatively minute mechanisms interact throughout the community in a kind of ripple effect upon stimulation by any outside influence. In a sense, homeostasis, if not the ecosystem at large, is in a constant state of perturbation in one constituent or another. Perturbation, however, is only recognized by the casual observer when a population, community, or whole ecosystem manifests symptoms of stress. By then, the cybernetic processes are, or have been, stressed beyond their ranges of tolerance.

PERTURBATION

Perturbation can be natural, or it can be anthropogenic. It is any disruptive intrusion into a biological system sufficient to overwhelm homeostasis or to impose a new homeostatic condition deemed detrimental, usually from some human perspective. For the environmental professional, the system most obviously of interest is the ecosystem. Today, we are beginning to comprehend that it is not enough to respond to perturbation in the name of healthful ecosystems.

Instead, the goal of the environmental professions is a healthy ecosystem. If the mission is not in the name of the ecosystem for its own sake, we have gained the wisdom to acknowledge that only healthy ecosystems can be healthful.

Chapter 10

Ecology of Aggregations

Among the features unique to life is the manner in which organisms are distributed. That distribution is essentially *never* random. "Never" is a word as infrequent in the vocabulary of scientists as it is in that of the lawyer. For that reason alone, aggregations are of particular interest to environmental professionals.

Working against the randomization otherwise imposed by entropy and the laws of thermodynamics, organisms occur in associations exhibiting interrelated structure and function, the influences of which are manifested at the ecosystem level. Thus, just as the cell and the entire individual organism represent levels of organization, so do populations of individuals of the same species and communities of populations characterizing ecosystems, culminating in the ultimate biological level of organization, the biosphere.

Throughout its life, the individual organism will occur in a habitat, serving in a niche in characteristic ways. The aggregation of organisms forming a population also occurs in characteristic patterns associated with habitat and niche, as does the aggregation of populations composing the community.

The resultant patterns can be studied and analyzed at any level of organization. They are better understood, of course, when the observer is aware of the contributing levels of organization. Moreover, all four dimensions must be taken into account. The patterns will change over the three dimensions of space and over time, the latter measured in fractions of seconds all the way to units of geological time.

It can be productive, then, to commence with either extreme, the individual or the whole community, as long as the holistic and reductionist approaches are both taken into account. Here we will begin with populations and build to the community. But then we will return to the individual in order to reexamine the ecology of adaptation and evolution.

Table. Statistical Functions of Populations and Corresponding Attributes of Individuals

Statistical Function	*Attribute of Individual*
Density	Existence
Natality	Birth
Mortality	Death
Age distribution	Age
Biotic potential	Reproductive activity
Dispersion	Physical location
Growth form	Life history

POPULATION ECOLOGY

A population is defined as an aggregation of organisms of the same species occupying a particular space over a given interval of time. For purposes of this definition, a species is a natural aggregation of organisms in which the individuals can exchange genetic information. In addition, a population is a biological system uniquely characterized by statistical functions and certain genetic characteristics. The statistical functions are density, natality, mortality, age distribution, biotic potential, dispersion, and growth form. The genetic characteristics are adaptiveness, reproductive (or Darwinian) fitness, and persistence (the probability of leaving descendants over long periods of time).

Statistical Functions

Since populations are composed of individual organisms and thus depend on those organisms for their features, it is informative to consider a population's statistical functions in comparison with the corresponding attributes of an individual.

The density of a population, based on the existence of the individuals composing that population, is the number of organisms per unit of space. Density depends on the interaction of the other statistical functions even as it itself contributes to them.

There will be a characteristic density for each population in an ecosystem against which actual density can be compared as a measure of the "health" of that population, with implications for the health of the larger community and the ecosystem itself. Most immediately, density interacts with natality and mortality, the population's birth rates and death rates, respectively. Natality is a phenomenon determined by a variety of factors inherent to the population and its component individuals and imposed by the environment in which they are interacting.

Inherent features are the age of sexual maturity and the theoretical number of young per birth and over the interval of reproductive activity characteristic of the species. These inherent features are modified by the actual age of sexual maturity among the individuals in combination with actual opportunities to breed and actual births.

Mortality is determined by comparable inherent and environmental features. Each population is characterized by an expected longevity among its individuals—the predicted maximum age at death. Individual life expectancy, however, actually alters in a characteristic pattern with the age of the individual organism. The actual mortality rate in the population thus depends on the actual age at death among individuals and is closely related to the cause of death.

Few animals die of old age, although plants and microorganisms may actually fulfill their life expectancies in the absence of perturbation. Similarly interconnected with natality and mortality and closely associated with density is the age distribution of a population, expressed as the age classes of the individuals. Age classes are made up of all the individuals in a specified age range.

Again, there are characteristic patterns to this statistical function that reveal the health of the population and reflect on the health of the community and ecosystem. An excess of aged individuals no longer capable of reproduction suggests the imminent demise of a population. Large numbers of young suggest a healthy population, but an excess would suggest the possibility of premature death among them, also an unhealthy sign. The age classes of the sexes may also be an indication of the prospects of the population.

Biotic potential is the maximum reproductive power of a population, unlikely to be realized in a healthy community. This function is closely related to the carrying capacity of the ecosystem but will be tempered by the population's interactions with other members of the community.

Dispersion of the population depends on the location of the individual organisms and the activities in which they are engaging. There are two levels of ecological factors determining activity and location. Habitat and niche as manifested as actual home ranges or territories of individuals are initially determinative.

These determinative factors, however, will be modified through the population's age distribution and depend on such time factors as the time of day and the season of the year. On the individual level, such additional factors as feeding habits, success in finding a mate, and success in producing offspring will all exert their effects on dispersion. Also significant are individual behavioral patterns that will be expressed within limits characteristic of the species.

Random distribution is at most rare. The other extreme—uniform distribution—suggests severe competition or antagonism. The most common pattern can be described as clumping. When not acting in isolation from their own kind, organisms of the same species tend to be distributed in pairs, clones, families, pods, schools, prides, or herds. Significantly, any attempt at statistical analysis of populations must accommodate, or "fit," the patterns assumed by the organisms themselves. Any lack of such fit will compromise analysis, with concomitant disruption of interpretations and conclusions.

Finally, each population manifests a growth form. The corresponding attribute of the individual is life history, which is predicted by species identity. The actual life history of the individual will depend on such factors as its prenatal environment; its actual development thereafter; and its progress through immaturity, sexual maturity, and senescence or old age. In a manner analogous to, and dependent on, the actual life histories of its members, a population grows, differentiates, maintains itself, and exhibits definite (and definitive) organization and structure.

While a population has no prenatal period, it is subject to development, and there are characteristic intervals of immaturity, maturity, and at least theoretically, senescence. Two major growth forms are recognized. Some populations exhibit a rapid exponential increase in numbers, coming to an abrupt halt as some limiting factor intervenes—the J-shaped growth form. Other populations exhibit an S-shaped form. Their numbers increase slowly before entering an exponential phase, which is followed by a gradual slowing.

Whether the expected form will occur will depend on all the features and factors, biotic as well as abiotic, influencing the population and its individual members. Growth forms can be disrupted by any number of factors, some of which can rise to the level of perturbation. As always, time is a crucial parameter. Groupings and patterns may undergo changes over specific intervals characteristic of the population.

From the foregoing points, it should be apparent that while absolute numbers of individuals in a population may indeed be important, changes in those numbers may be far more informative. In the first instance, there are both theoretical and actual upper and lower limits on population size itself. Too few organisms suggests the possible imminence of extinction.

The upper limit on population density depends on environmental conditions. Most significant among the external conditions are energy flow and the trophic levels. Internal conditions, such as the size of the organism and its rate of metabolism, are also significant to the upper limit on population numbers.

Too many members subjects the individuals and the population itself to a variety of stresses, which in the extreme can be lethal. At sublethal levels, stresses can stimulate responses ranging from those promoting survival of the species to those promoting detrimental abnormalities. Some organisms (including the Chincoteague ponies made famous by author Marguerite Henry) become stunted. Smaller organisms require less energy and less food than their normal counterparts.

In others (including at least one species of deer), the female's body resorbs the developing fetus. Resorption can be deemed a positive response because the female avoids the stress imposed by the demands of a growing fetus. One potential member of the population is lost, but perpetuation of the species is advanced by the female's survival and the increased likelihood of additional offspring under more tolerable conditions in the future.

In some populations (including laboratory populations of rats), crowding results in abnormal Behaviour that can include cannibalism among the adults and of newborn litters by their mothers.

GENETIC CHARACTERISTICS

The gene pool of a population will support a characteristic range of phenotypes initially manifested at the molecular level and then throughout the ranges of features defining the species. These features range from the biochemical and physiological to behavioral. Each feature, as well as every response to limiting factors and every tolerance range, will be subject to a range unique to that species or that particular population within a larger species. Tolerances are genetically determined and passed on from generation to generation, subject to alterations in the genetic material, the genome of individuals, and the genetic pool of the population.

Because population genetics is statistically dependent on the genomes of the individual members of the population, two ranges actually exist. One is the range characteristic of the population, in which most of the members will fall.

The other is that effected by the actual genome of the individuals. There will be occasional individuals who are beyond the minimum and maximum for their population, and this outsider status may be detrimental or beneficial to the individual.

The detriments arise when the individuals cannot tolerate a condition in which its fellow organisms survive even when they do not thrive. The benefits arise when individuals can survive or even thrive in conditions beyond either extreme. The capacity to meet environmental contingencies can ultimately lead to a more resilient population or to actual speciation.

Significantly, at least simple populations can be studied under controlled laboratory conditions. Here, group attributes lend themselves to measurement, and therefore environmental factors can be imposed and the results on those attributes described in qualitative and quantitative terms. Such relationships are not readily isolated for clear analysis in natural populations.

THE ROLE OF BEHAVIOUR

In earlier discussions, the importance of the third spatial dimension, height, in addition to the two horizontal dimensions, was emphasized. Similarly, the importance of the fourth dimension, time, has been repeatedly emphasized. Neither ecology nor the practice of the various environmental sciences is complete without those two dimensions. They are ignored at the risk of the quality of analysis and, where a political or legal decision affecting the ecosystem rests on that analysis, at the peril of the ecosystem.

Another "dimension" characterizing the biota is equally important yet subject to neglect: Behaviour. Behaviour is a complex of the phenotype and is just as dependent on the genome as more easily studied features. Behaviour is not limited to animals, although in animals it is complex to the point of limited predictability, especially among species with highly developed brains. Behaviour is an attribute of the simplest organism, and the highly developed animal brain is nothing more or less than the other extreme of a continuum.

Behaviour is an attribute of all organisms and occurs on

a continuum ranging from tropisms of plants to the profound ability to reason that is attributed to humans and to other organisms humans consider to be their companions high on the scale of complexity approaching intellect. In fact, these components of Behaviour tend to blend from one to the next, with unique leaps in taxons sporting central nervous systems and boasting the next step, an organized brain. Tropisms are readily observed in the higher plants. A tropism is a movement or orientation mediated by hormones. Plants exhibit three kinds of tropism. Geotropism causes roots to grow downward into the soil, and the remainder of the plant to grow upward. Heliotropism—particularly apparent in sunflowers—and phototropism cause herbaceous plants to move with the sun and with light, respectively.

Taxes are manifested by single-celled organisms and by organisms as complex as insects. A taxis is a directed reaction to a physical or chemical stimulus. Different kinds of protozoa move toward, away from, or in a direction transverse to a chemical stimulus or a physical one such as light. Moths are drawn to light, and bees orient themselves under the stimulus of polarized light.

Both tropisms and taxes are innate to the organisms manifesting them and may or may not require a central nervous system. Also innate are reflexes, but here, cells and tissues must be organized into organs or organ systems. One familiar reflex is salivation at the sight or, more intensely, the scent of food, a reflex that can be conditioned, a simple form of learning. Another familiar reflex is the (original) knee-jerk reaction—the sudden, uncontrollable snap of the lower leg when a sharp tap is administered just below the bent and relaxed knee. (The figurative knee-jerk reaction associated with political affairs is presumably conditioned but requires no well-developed brain.)

A more subtle reflex is the spontaneous removal of the hand from a surface that might be hot. This reflex, about which the "victim" may grin sheepishly with a shake of the head upon "remembering" that the stove or iron or whatever was not on at the time, is identified as a spinal reaction. The stimulus need not reach the brain; the reflex is mediated by nerves in the spine. There is actually a selective advantage to the speed associated with such reflexes.

The components of Behaviour identified as instinct, learning, and reasoning are subject to substantial controversy. While scientists prefer to label such phenomena and then place actual manifestation of them neatly in the labeled categories, realities are not so amenable to labels or categories. Instincts, which are stereotyped Behaviour patterns fixed by the gene pool of the taxon and essentially immutable once stimulated, are actually complex patterns of Behaviour that are subject to some level of learning determined in some part by the complexity of the nervous system. The distinction between instinctual and learned Behaviour is not an easy one.

There are those who will argue that human Behaviour is all learned without resort to any instincts. But there are others who maintain that there are human instincts. Those who recognize human instincts include these innate, programmed responses among human Behaviour patterns. Here, there is special reference to Behaviour associated with primal needs, notably including thirst, hunger, and sexual drive. Of all the organisms, humans are the most likely to temper both learned and any instinctual Behaviour with reasoning.

All of these intraspecific interconnections are simultaneously participating in those occurring among and across populations to give rise to the community.

COMMUNITY ECOLOGY

Aggregations and even associations among individuals of a population could be no more than fortuitous, but ecological research has revealed interactions causing population patterns. If, like the organism and the population, a community is a legitimate biological system and identifiable level of organization, what questions are appropriately applied to it? The first and most obvious question is, What identifies an aggregation of populations as a community, an *association* among populations?

How is a community organized, and how do its

component populations interact? These questions raise the next one: How is a community effectively analyzed? Finally, what does community analysis reveal of the ecosystem, and to what extent can the revelations be applied to ecosystems generally? What do they tell us of the biosphere? For the environmental professional, what are the practical implications of community ecology?

At the simplest level, the community is the living component of an ecosystem. A community is an organized assemblage or association of populations in a prescribed area or a specific physical habitat. While assemblages can be no more than fortuitous with no associations among organisms, a community is defined by physical, chemical, and biological connections. Community or ecosystem boundaries may be sharply defined and separated from others, or one may gradually blend into another.

The most immediate question is the extent to which a community is more real than an artificial construct merely serving the convenience of the observer. That lines are no more simply drawn at this level of organization than elsewhere is reflected in the "closed" and "open" communities of ecologists. These designations are a reflection of interconnections across whatever lines may be drawn around ecosystems.

Every ecosystem is intimately connected with at least contiguous ones, and all are ultimately connected with the biosphere. In attempting to draw lines distinguishing the real from the artificial, there is one governing principle: The efficiency and stability manifested in a community of organisms increase in direct proportion to the degree of evolutionary adjustment marking relations among populations. An aggregation of organisms among which there are highly organized interactions giving rise to perceivable efficiency and homeostasis at least suggests the existence of a legitimate level of organization—in short, a biological system.

Relationships within a community can be described in terms of correlation. A mere aggregation of randomly distributed organisms will reveal zero correlation. Relationships, on the other hand, are described in terms of positive or negative correlations and in terms of the strength of either correlation. The most extreme evolutionary adjustment is coevolution to the point of obligate association in which pairings between species or other combinations represent mutual interdependence.

This mutualism includes a range of interactions. Certain flowers are pollinated by certain insects, and fruits may be distributed by herbivores for later germination. More intimately, lichens are an obligate combination of an alga and a fungus. Ruminants and termites depend on internal microorganisms capable of digesting cellulose. In their roots, legumes support populations of the bacterial genus *Rhizobium*, whose species fix nitrogen for biological assimilation and distribution.

Significantly, patterns of mutualism probably evolved from antagonisms such as parasite and host, predator and prey, or plant and herbivore. The opposite extreme is independence, a most unlikely situation in nature where individuals of each species would be randomly distributed with respect to those of other species. Between the two extremes are a variety of patterns of association.

COMMUNITY ORGANIZATION

A community is an organized unit at least to the extent it exhibits characteristics above and beyond those of its individual and population memberships. Like those biological and ecological units, a community functions as a unit through coupled metabolic transformations. In the case of the community, those transformations are based in the effect on populations of products of metabolism and catabolism of others in addition to the more obvious direct interactions at the individual and population levels.

Major communities are so characterized on the basis of sufficient size and completeness of organization to allow for relative independence. Major communities need only the externality of the sun's energy for support. They are relatively independent of the resources (outputs from) and requirements (inputs to) of adjacent communities. Minor communities, then, are characterized by a greater dependence on neighboring aggregations. Each community is so organized as to manifest definitive functional unity, based in a characteristic trophic structure and associated patterns of energy flow. There is some compositional unity, that is, a definitive array of taxons. In other words, there is an expectation of the presence together of certain organisms, but species are to some extent replaceable by others over space and time in the ecosystem. The niches, however, are relatively fixed and open to a limited array of organisms.

The community defining a specific ecosystem, whether a seral stage or a climax community, starts with the abiotic environment. Whether that community represents a seral stage, the climatic climax, or an edaphic climax will be determined in the first instance by environmental gradients in such familiar features as latitude, altitude, and climate, both regional and local.

Between ecosystems, unique communities may also exist. The ecotone represents a transition habitat separating one ecosystem from another. An ecotone community may include organisms (possibly distinct ecotypes) from adjoining systems and characteristic populations that may be limited to niches found only in such transitional settings.

INTERSPECIES RELATIONSHIPS AND COMMUNITY ANALYSIS

Interspecies relationships in a community may be sharply defined and separated from others or may blend into one another. The overall result of interactions within a community can, as has already been described in some detail, be measured and qualified in terms of standing crop as well as in terms of overall community metabolism. Since the most conservative estimate would acknowledge the presence of hundreds of thousands of individual organisms in a community, focal points of analysis are essential.

In selecting such points of departure for purposes of analysis and comprehension, the assumption is that relatively few of those organisms exert the major controlling influence through their numbers, size, or production. Moreover, representatives of those few are enough to provide meaningful information for purposes of ecology or the environmental sciences. This, after all, is the major premise behind statistics and the applications of biostatistics in science and law.

The observer seeks out ecological dominance and species diversity to characterize a community. Ecological dominance is categorized as the concentration of influence in one, several, or many species. Indices of dominance are based in major structural features, physical habitat, or functional attributes. Species diversity is measured in terms of ratios between the number of species or individuals in a species present and the total number of individuals present.

Higher diversity is associated with longer food chains, more symbiosis among species, and greater feedback control, all of which serve ecosystem homeostasis by reducing oscillations and increasing stability among and within populations. Diversity is believed to be high in older communities and low in newly established ones.

COMMUNITY DEVELOPMENT

The development of an individual community will follow the strategy of ecosystems previously discussed. Still, every community will be unique within genetic and abiotic constraints. Moreover, each community will be the result of any perturbations afflicting the ecosystem directly or indirectly.

Other than the matter of inherent value of this biological system, what are the practical implications of communities and their respective organizations? The answers lie in the aphorism, As the community goes, so goes the organism, and its numerous corollaries.

LESSONS FOR THE ENVIRONMENTAL PROFESSIONAL

Environmental professionals can learn much from what research ecologists have been asking of ecosystems in recent years, starting with an ecologist's view of the minimum requirements for meaningful experiments into community components as elsewhere. They are threefold: knowledge of initial conditions, adequate controls, and replication. For both researchers and environmental professionals addressing questions raised in a politicolegal context, the most common failing in studies lies with inadequate descriptions of the preexperimental situation. Significantly and somewhat ominously, this failure cannot be overcome without a sufficient period dedicated to the collection of baseline data, the status quo ante.

Similarly, controls are a ubiquitous problem but most particularly for the applied ecology practiced by most environmental professionals. Under most circumstances, a discharge or other pollution event will be into a receiving environmental medium (an ecosystem or portion thereof) for which there simply is no control. In lieu of more traditional controls, an effort is made to work from relevant literature to project what would have happened in that ecosystem absent the anthropogenic intrusion. In the absence of acceptable controls, similar conjectures must be made in deciding whether to attribute any alteration in the ecosystem to the intrusion in question.

These sorts of problems can be clarified with reference to microcosms representing population or community ecology. For example, the role of controls is clarified in an experiment to determine whether two species in the same genus are competing in a habitat. In seeking an answer, the researcher must set up experimental plots distributed throughout the habitat.

In one series of plots, one species will be removed in the expectation that the other will increase if the hypothesized competition is in effect. In a second series of plots, the other species will be removed in the same expectation. In a third series, neither species will be removed.

Thus, if one species does indeed increase in its experimental plots, the effect of competition on that species is demonstrated. Or is it? What if a comparable increase occurs in the control plots? The control plot is only one control here.

Equally important is the fact that each condition, the control and two experimental ones, is the subject of more than one plot. This is an effort to accommodate any subtle differences imposed from place to place in the habitat as identified. It is also essential for purposes of internal replication.

Replication is yet another obstacle to sound experimentation for ecologists and environmental scientists. It is literally impossible to find identical locations for experimental comparisons because of the pervasive variability in natural systems. The resultant uncertainty is whether unrecognized systematic differences are present among the locations.

Replication of the application of experimental conditions is at once a kind of control and internal replication of the experiment. A second and equally important kind of replication is that of the experiment itself in the same habitat, elsewhere, and by the same or different researchers. In designing research programs, research ecologists have publication in mind, while most environmental professionals have missions with externally imposed deadlines to meet. There may be a crucial difference here:

The researcher may be motivated simply to understand the subject system and will manipulate it for that purpose, while the environmental professional may be motivated or constrained by factors that are not compatible with the realities of the system. One often hears the argument that the information sought is crucially important and must be put to immediate use in the marketplace or under the law.

The usually unstated premise is that anything is therefore better than nothing. But that premise is highly suspect. If the information is critical, time is not the sole factor to be considered. Certainly, the accuracy of the information is equally essential, if not more so.Despite the complexity of real ecological systems in comparison to their laboratory counterparts, there is considerable pressure to devise ways to study natural systems for the reason that their very complexity makes them far more real.

They will provide a more nearly "real" answer to the

question at hand. A related practical problem is the matter of explaining the result that fails to substantiate the underlying hypothesis. There are no less than four potential explanations for failure to confirm a hypothesis in a natural system. All are worthy of contemplation by the environmental professional.

- The expected phenomenon is irregular in occurrence.
- The experimental design was inadequate.
- Experimental execution was inadequate.
- The hypothesis is wrong.

It is worth remembering here that almost all of science, including ecology and the environmental sciences, is a matter of inference. The first and foremost inference is that events are not purely random. There is a causeand-effect relationship in some form or in a variety of interactions. Scientists often use the Null Hypothesis to advance theories. By this process, the disproof of a variety of alternative explanations of an event or phenomenon serves to advance the "preferred" hypothesis.

The process is necessary because logically one cannot prove the premise by the occurrence of the expected conclusion; one can only disprove the premise by the failure of the expected conclusion. With this limitation in mind, scientists go about reinforcing hypotheses by approaching the hypothesized premise from a variety of intellectual directions.Statistical probabilities are a kind of Null Hypothesis used to establish that a phenomenon has not occurred at random or by chance and is instead more likely to be somehow related to a specified phenomenon as predicted.

Thus, to be as complete as possible, a researcher must show at least three things:

- The phenomenon could be causally related to the hypothesized premise.
- The phenomenon has not occurred by chance.
- The phenomenon is not attributable to some other cause (including one or more causes of the hypothesized premise).

A subtle difficulty arises whenever an experiment is performed. To what extent is the result attributable to or altered by the experimental technique itself? Notorious examples are "cage effects," whereby isolating the experimental system itself affects the system positively or negatively.

Specific situations include the ability of predators to follow investigators' trails to the subject (prey) organism and the existence of "trap-happy" individuals in a population. These are organisms capable of learning that a desired object is available in the trap without expenditure of energy to search it out elsewhere and without risk of harm.

With particular reference to community ecology, the conventional wisdom is that the greater the diversity, the healthier and more stable the ecosystem. Is this conventional wisdom supported with experimental results? The hypothesis is that as species are added to the community in the course of the sere, each takes up a unique portion of the niches that have been established by successional predecessors. Thus is each niche enriched, and the loss of a portion of it need not spell doom for related niches and eventually the ecosystem itself. The fluctuations to which earlier invaders were subject are reduced, and homeostasis is protected.

The question to be raised through appropriate experiments is whether there is any evidence to the effect that fluctuations are any less dangerous to small populations than to large (and what the answer means to the ecosystem). The related hypothesis is that smaller populations are at greater risk of extinction than larger ones. Can the competing hypotheses be reconciled?

How are causes and effects to be sorted out in this situation? Is it that stability enhances diversity? Are there points in a sere where cause and effect actually reverse? Is this a matter of positive feedback? Is positive feedback of indefinite duration, or can there be a point at which feedback turns from positive to negative? Such questions are not subject to immediate answers, for all their significance to the research ecologist and the environmental professional.

Of equal importance to both is the matter of diversity. It has been observed that a pattern of abundance exists in nature. In any given area, a few species are abundant, while many are rare. An unfulfilled mission of ecological research is to devise a formula for a statistical model or a statement of that phenomenon for the purpose of guiding comparison among communities. Any "index of diversity" would be a declaration of the richness of the community.

The results would represent hypothetical community organizations against which actual communities could be compared for fit. How can matters or issues before the environmental professional that turn on diversity be addressed or resolved in the absence of such models?

Overall, there are instructive conclusions to be drawn from ecological experimentation in the field. Generally, however, a major caveat holds: "Each kind of environment should be considered separately, because there are few, if any, specific statements about ecological processes that will be true across all environments." The great coherence among conclusions is associated with forest communities.

There, experiments within trophic levels yield consistent results, and relationships between trophic levels are consistent with those within. For example, it has been shown that interspecific competition is significant among decomposers. Negative interactions occur among species in almost all systems, and interference competition is consistently demonstrated. In the grazing cycle, distribution of herbaceous species is determined by distribution of nutrients or toxic ions. Trees adapt to herbivore damage, but adaptations to those defenses by herbivores is both effective and ubiquitous.

In successional communities, relationships within and between most trophic levels are less consistent than in forests. Decomposers are an exception in that they are always in competition. A related phenomenon is that "accumulation of organic matter in most terrestrial situations is too slow to be of importance in preventing competition among the fungi and bacteria." There is also some suggestion of mutualisms: "In the process of chewing the dead plant matter, the detritus feeders create smaller particles and thus larger surface areas for the true decomposers to colonize."

A series of conclusions of some relevance to the applied

ecology of environmental professionals have followed from research among successional plants:

- Competition is sometimes present.
- When competition is present, it is weak or diffuse.
- Some studies show that herbivores have an important impact.
- At least minimal competition occurs among herbivores.
- Predation has a significant effect on populations of some herbivores.
- Competition among predators is difficult to demonstrate.

Because when the exigencies of the marketplace and of the politicolegal system are imposed on this kind of natural conundrum far more questions are raised than answered, the environmental professional may well cringe in the recognition that the legal profession is not the only one to whom "weasel-words" are attributed.

The foregoing questions as presented are hypothetical ones. But they can be of critical importance in the professional settings of environmental practitioners. Yet the questions are not answered in this chapter or anywhere in this volume. Instead, the principles and discussions provide the tools by which the answers and others like them can be sought effectively in terms of scientific ecology, presumably the foundation of environmental protection in all its forms.

There are two relevant observations to be heeded by the environmental professional: As the organism goes, so goes the community, *and* as the community goes, so goes the organism. Hence, in order to "manage"-encourage or discourage—a specific organism, it may be more effective to manipulate the community or some portion of it rather than the target organism itself.

ECOLOGY, ADAPTATION, AND EVOLUTION

Ecological successes and failures are related to diversity among organisms, adaptation of and among biological systems, evolution, and extinction. Primary inquiry into these interrelationships seeks mediating mechanisms. Ecology and evolution are related through the concept of species as expressed through populations. What happens when a single species is introduced to an area with open niches? What happens when there are comparable open niches in physically separated regions?

ADAPTATION, ECOLOGICAL AND EVOLUTIONARY

Adaptability commonly incorporates two concepts, for simplicity, distinguished as ecological and evolutionary adaptations. Throughout its life, an individual adapts to an array of environmental conditions or suffers the consequences of failing to do so. Such an *ecological* adaptation is founded in the genotype that individual has inherited. Similarly, populations, whole communities, and even the larger ecosystem are constantly adapting to environmental circumstances. None of these ecological adaptations calls for alteration in the gene pools of the affected biotic aggregations. All are relatively temporary.

But adaptation is also used in a frankly evolutionary sense applicable only at the population or community level, despite necessarily being expressed through outstanding individuals. At any given point in time, living species and higher taxons reflect adaptations to enduring environmental circumstances. These evolutionary adaptations have been mediated by genetic alterations allowing for survival of generations of organisms and becoming fixed in gene pools at least for the duration of the corresponding selective pressure. In contrast to the purely ecological sense of the word, evolutionary adaptation tends to be permanent, fixed in the genetic code.

There is a relationship between these two kinds of adaptation, expressed through individuals. In the first instance, evolutionary adaptations include individual ability to adapt to constantly altering circumstances. More important for future generations, if a population encounters environmental challenges (selective pressures) exceeding its range of tolerance, individuals limited to existing ecological adaptations will be eliminated along with their genes. Only those members genetically endowed for tolerance beyond the extreme will survive to produce offspring bearing their genes. These offspring will carry that genetic endowment into the future. The number of affected individuals will increase relative to the rest of the population until the more tolerant individuals compose the whole of the population. The population has adapted in the evolutionary sense and may have undergone speciation in the process.

The surviving species is adapted to a new niche. Evolutionary adaptation has resulted in ecological adaptation. In both cases, homeostasis is the immediate strategy. Significantly, the strategy of the species is survival over generations rather than survival of either the individual or of even a single generation.

Even though adaptation in both senses may be expressed at any number of levels of organization, the result is based in molecular responses. Examples include alterations in the structure of enzymes when temperatures are high, the breakdown of photosynthetic pigments in excessive light, and altered configurations in protein molecules with variations in salt content of the environment. Every adaptation requires energy.

Once again, ecology comes full circle, linking the holistic to the reductionist. Synecology and autecology are not severable. In order to begin answering the opening and related questions, it is necessary to return to a reductionist approach. It is necessary to consider the individual organism and the species in the ecosystem.

Genetics is an informative point of departure for establishing linkages among ecology, adaptation, and evolution. Adaptation and evolution represent ecological success, while extinction represents failure. Genetics can explain both.

THE SCIENCES OF GENETICS

All living beings, extant as well as extinct, share two features: All are based on the chemistry of carbon, and all share one of two specific molecules in common—DNA and RNA.

These genetic molecules, by which traits are passed on from generation to generation, paradoxically represent both the unity and the diversity of life.

GENETICS OF THE INDIVIDUAL

Genotype and Phenotype

The individual organism is the expression, phenotype, of the genotype, the genetic basis of identity and function of all organisms. Each individual is the expression of the code incorporated in its genome. That genome is unique to the individual within the constraints of the taxons to which it belongs. The species is defined genetically by the number and structure of its chromosomes, the bodies on which the genes are located.

At the molecular level, those genes are an expression of the order in which four bases appear along the strands of DNA making up each individual chromosome.

The process by which phenotype emerges from genotype can be summarized as follows:

DNA → RNA → Amino acid Sequences → Proteins

Genes are expressed in the somatic (body) cells of multicellular organisms.

In actuality, a specific complement of genes, not necessarily on the same chromosome, will be expressed in an elaborate pattern of protein synthesis with the exquisite timing required for the cell to function properly as a whole. If the organism is a single cell or an association of like cells, coordination and timing need go no further. In multicellular organisms, especially those composed of interacting tissues, organs, or whole organ systems, the expression of individual gene complexes acting together will be serving the needs of the whole organism.

Obviously, there is far more to an organism as it functions in the ecosystem than has yet been fully revealed in the above sequence; and in fact, the complex of molecular events lies far outside the scope of this discussion. Suffice it to say that the basis of the complexity we know as life is ultimately the result of individual genes switching on and off, with consequent molecular changes occurring intracellularly. Those molecular changes are very precisely integrated for expression at the cellular, extracellular, and organismal levels.

Moreover, in the somatic cells, there is a precise division of Labour supporting in these multicellular organisms the structure and function of tissues, organs, and organ systems, all the way up to the organism itself. As will be discussed below, this division of Labour is the basis of physiology, taxonomy, ethology, and both ecological and evolutionary adaptation.

Sexual Reproduction

Among organisms that reproduce asexually by the simple fission of cells into daughter cells, there is limited opportunity for genetic variability. Instead, in bacteria, for example, genes can be turned on and off to meet the exigencies of the environment, with special reference to altering chemical content. Bacteria adapt readily to altering nutrient supplies as well as to an altering toxic environment.

Moreover, these and other organisms undergoing asexual reproduction tend to have very short generation times. Rapid division alone can allow for the distribution of new phenotypes from generation to generation. But even bacteria occasionally exchange genetic information between organisms in a kind of forerunner of sexual reproduction.

This exchange of information allows for enhanced variability that tends to expand the environmental horizons of those who participate.

Sexual reproduction is a considerable advancement, if only in terms of genetic variability and environmental adaptability. Most members of the plant and animal kingdoms engage in sexual reproduction and possess specialized gametogenic cells in addition to their somatic, or body, cells.

If each organism is the subject of a specific and unique genotype passed on through all its generations, how is the genotype distributed? How is it that the chromosomes do not all pile up and crowd out all the rest of the intracellular components?

Cytogenetics

Molecular genetics, molecular biology, and cellular physiology are all "visible." The visibility is the subject of the science of cytogenetics. Cytogenetics has revealed that at certain intervals each cell of an organism provides a microscopic picture of genetic processes. Upon division, the chromosomes in each cell "materialize" and proceed in a "dance" of extraordinary precision.

Each somatic cell contains the diploid number of chromosomes characteristic of the species. The diploid number for humans is forty-six, or twenty-three pairs. Each parent contributes one set of the paired chromosomes. Moreover, there is a karyotype; that is, the chromosomes have shapes and sizes even more precisely characteristic of the species than the diploid number alone.

Each chromosome also has a characteristic structure, including a centromere, a "swelling" that occurs at a specific location along the length of the chromosome. Cytogeneticists can identify an organism by its karyotype and even find certain kinds of gross genetic abnormalities represented by extra or missing chromosomes or misshapen ones.

Every time a somatic cell divides, it first goes through the process of mitosis. Within the nucleus, each chromosome has been duplicated during interphase, when the chromosomes are diffusely distributed as chromatin. Just before cell division, the doubled chromosomes materialize and begin to line up end to end in the centre of the nucleus. In humans, all forty-six doubled chromosomes, each consisting of two chromatids, line up at random, and then each chromatid is drawn away from its partner by its centromere.

Simultaneously, the nuclear membrane commences to indent and causes a split in the nucleus between the separating chromatids until the two sets of forty-six chromosomes are fully separated. Both the nuclear membrane and the cellular membrane pinch off to result in two new ("daughter") cells, each with an intact cellular membrane and a nucleus in which the chromatids fade into the dispersed chromatin of interphase.

The forty-six chromosomes will each again be duplicated into two chromatids during the subsequent interphase. Thus, under normal circumstances, each cell of a multicellular organism has a complete set of the genetic information identifying that organism. But any mistakes in the code will also be duplicated in subsequent generations of cells.

If each parent is to contribute only half the genome to offspring, a mechanism must exist to accomplish that reduction in the gametes, or germ cells—the sperm and ova of multicellular organisms.

The mechanism is the process of meiosis, which generates haploid gametes. Like their somatic counterparts, germ cells go through a nuclear interphase during which, using the human example again, the forty-six chromosomes are doubled into chromatids connected by their centromeres. In the course of meiosis, two cellular divisions will occur in rapid succession.

This time, however, the chromosome from one parent will pair with its counterpart from the other parent. The result in the first meiotic division is the lining up of paired chromosomes so that there are four chromatids aligned lengthwise in the centre of the nucleus. Here, the forty-six chromosomes are randomly distributed along the line in twenty-three pairs; each member of the pair, consisting of two chromatids, is separated by its centromere.

Each of the two daughter cells has a full complement of fortysix chromosomes, each consisting of the two chromatids. Those from each parent, however, have been randomly distributed. Some of the forty-six are from one parent, others from the other parent.

Now, instead of entering interphase, a second cellular division occurs. The forty-six chromosomes of mixed parentage again line up in the two daughter cells, and the two chromatids separate, resulting in a total of four gametes, each with the haploid number of twenty-three chromosomes. When the gametes combine to initiate development of the offspring, the two sets of the halpoid karyotype are reunited in the diploid cell, which proceeds through division and differentiation to form the new organism.

This whole process allows for variation attributable to genetic differences in the parents and still greater variation through mutation and exchanges of genetic information between paired and doubled chromosomes. Such variation accounts. For individual differences within a species and can also account for failures of some young to survive and for speciation when such an alteration meets the needs of selective pressures in the environment.

POPULATION GENETICS

Genetic variation among its individuals represents the genetic content of the species, the population's gene pool. This gene pool is a qualitative and quantitative description of the population. A kind of biological "law" holds that under specified circumstances the gene pool will not be altered but will remain stable from generation to generation.

This Hardy-Weinberg Equilibrium is a mathematical description of how genes, including any that have mutated or otherwise been modified, are distributed in the population. Once established, the percentage of individuals expressing a new gene will stabilize and thereafter remain constant over any number of subsequent generations.

The Hardy-Weinberg Equilibrium thus describes a population that is undergoing no quantitative alteration in ratios among genes in its gene pool. The ratios of individuals carrying any given genotype will remain constant.

Neither negative nor positive effects would influence the ecological status of the population. But for the stability of this theoretical equilibrium to be realized, several factors must be in operation:

- The population must be closed, without gains from or losses to any other population.
- (Additional) mutations either do not occur or do not

- Disturb the established ratio of genotypes.
- Reproduction is at random.
- The population is large enough that chance alterations in frequency are insignificant.
- No natural selection is occurring.

One need be neither geneticist nor ecologist to be aware that this particular array of conditions is highly unlikely in any ecosystem, even assuming it is closed to immigration or emigration. Mutations, harmful and beneficial, will be occurring spontaneously.

Moreover, their effect may alter substantially as alterations in the ecosystem occur. An inherited feature that was neutral or detrimental can become beneficial, and all kinds of permutations or changes in environmental circumstance can occur as the ecosystem undergoes natural alterations or perturbation.

All the array of ecological interactions will impinge on the gene pool and the implications of any given mutation transform accordingly, with corresponding alterations among ratios. Reproduction is almost certainly never at random, and indeed, mate selection and reproductive success from individual to individual will be major factors in the status of the gene pool and the population.

Adaptation, speciation, and evolution itself will all depend on this reality. Natural selection will be present to some degree.

TAXONOMY AND REALITY

Clearly, ecology of both the individual and the species depends on, or at least can be explained in terms of, physiological processes in cells and whole organisms. Molecular biology, once the antithesis of ecology, is now recognized as a source for ecological events and for the structure and function of the ecosystem.

It is truly astonishing that the basis of inheritance among all of life in all its possible diversity of life—extant, extinct, and yet to evolve—is mediated by four simple molecular entities, the bases of the DNA molecule.

Table: Organization of Living Beings—A Typical Example

Taxon	*Identification*
Kingdom	Animalia
Phylum	Chordata
	(Vertebrata)
Class	Tetrapoda
	(Mammalia)
Order	Primata
Family	Hominidae
Genus	Homo
Species	sapien
	(sapiens)

From photosynthesis to respiration to the buildup and distribution of biological chemicals to their catabolism and elimination, cellular physiology is at work. Even death represents physiological processes that not only define the boundary between life and death but also participate in biogeochemical cycles.

The ecologist, then, cannot safely ignore the detail of biochemistry, which parallels the wealth of detail encountered at the ecosystem level. Even adaptation, evolutionary as well as ecological, is ultimately mediated, or at least explicable, at the molecular level. Here is a true emergence of phenomenal order out of complexity beyond most imagination. Taxonomy is the biological science charged with extracting the order from the chaos of complexity. The interconnections are expressed and utilized in reconciling reality with perceived taxonomic relationships.

Taxonomy seeks to establish what might be called familial relationships among living organisms. Taxonomy also seeks to find family lines linking existing organisms to their evolutionary pasts and to reveal familial relationships among organisms known only from the fossil record. Taxonomy is also responsible for identifying every organism with an array of taxons from kingdom to species and subspecies.

Note that there is a structure here, and it is not far removed from the Theory of Integrative Levels. As taxons move from kingdom to species, each step is more specific than the last. But, at the same time, that which defines the higher taxon is retained throughout the lower.

The human being has been defined, not entirely facetiously as a featherless biped. In taxonomic terms, *Homo sapiens sapiens* is a member of the animal kingdom, distinguished today not only from the plant kingdom but also from Monera (bacteria), Protista, and fungi. Animals are multicellular organisms characterized by a cellular membrane but lacking the cell wall of plants. Animals are incapable of photosynthesis, but all of them respire.

Homo sapiens sapiens is among those animals with a chord of nerves protected by the bony structure of a backbone. This phylum places humans among the animals that are bilaterally symmetrical. Humans are vertebrates, a subphylum of all chordates. Humans have four limbs; they are tetrapods. As mammals, they are warmblooded, bear live young, and provide nourishment for the newborn by secretions of the mammary glands.

Most have hair, a specialization of cells in the organ called skin. Humans are primates; their limbs are specialized in two sets for locomotion and for grasping, respectively, and they possess particularly large brains of a unique configuration giving rise to high intelligence.

There is one extant genus of human, *Homo*. There is also one extant species, *sapiens*. That extant species is distinguished from all its predecessors as the subspecies identification, the second *sapiens*. We distinguish ourselves from all other primates and any other species in our genus in terms of our intellect, especially as manifested in the cultural organization of our populations.

There was a time when taxonomy was held by some, especially nontaxonomists, to be entirely artificial, a mere convenience in aid of communication among biological scientists. Based essentially on morphology and other relatively obvious similarities and differences in the outward appearance and some

Behaviour patterns, any correlation between taxonomic identification and actual relationships among taxons was dismissed as fortuitous. Now, those obvious criteria are accompanied by distribution of biochemical pathways, relationships among karyotypes, and actual molecular structure of DNA in efforts to sort out members of the biota into taxons reflecting evolutionary accuracy.

CURRENT EVOLUTIONARY THEORY

NEO-DARWINISM

Neo-Darwinism explains Darwin's concepts of survival of the fittest and origin of species in terms of the full array of biological sciences, including ecology. Ecological and evolutionary adaptation constitute a crucial component of neo-Darwinism, which now recognizes genetics, mutation, and population genetics as the mechanisms of evolution. The preceding discussions of individual and population genetics represent the underlying foundation.

PUNCTUATED EVOLUTION

The current theory of the extinction and origin of species goes beyond neo-Darwinism in recognizing that the slow processes associated with population genetics cannot fully explain the fossil record. This slow process is "punctuated" with sudden losses of major taxons (most notoriously, the dinosaurs) and unusually rapid emergences of others. These periods of punctuation are attributed to a variety of phenomena not far removed from cosmology and the sciences associated with comprehending our galaxy, solar system, and planet. Once again, the natural sciences come full circle as evolution affects ecology and is explained in terms of phenomena beyond the global.

EVOLUTION AND ECOLOGICAL NICHE

How organisms adapt and evolve to fill ecological niches is pieced together from a variety of sources almost exclusively by inference. Inferences are derived from interpretations of the entirely empirical sciences of paleontology in combination with current field studies, in which some experimentation can clarify empirical findings. Some laboratory analyses can also contribute to relevant inferences.

The pattern is perceived as one of rapid diversification among organisms with concomitant geographical expansion of their ranges and territories. With that expansion comes adaptive radiation—"formation of two or more new species, macromolecules, or physiological pathways, adapted to different ways of life, from one ancestral species, molecule, or pathway"-followed by internal extinctions, with the survivors becoming highly specialized.

Examples of the diversification stage are familiar ones. Insects and microorganisms fill every known habitat. Reptiles invaded terrestrial habitats, marsupials invaded Australia, mammals invaded the whole of the biosphere with the extinction of large reptiles, and birds invaded the whole globe with their ability to fly. In each case, diversification followed closely upon the invasion.

Adaptive radiation is followed by intervals of little evolutionary change. Extinct species will be replaced by close taxonomic relatives. Such evolutionary conservatism is advantageous—until the next important perturbation. Thereupon the process is reinitiated: adaptation and evolution or extinction.

APPLICATIONS IN THE ENVIRONMENTAL PROFESSIONS

Whenever we step into an ecosystem, we are taking a chance and may be misguided even before we take action:

We tend to think of familiar natural landscapes as frozen in time, like the image in a photograph, to remain unchanged. To preserve the area as we see it, we protect it against fire, insect attack, and other events we consider harmful. In doing so we change it, because we fail to comprehend that nature is not constant, that disturbance is the means by which landscape diversity is maintained.

In the end, "an environmentally induced modification of

a character is not inherited. What is inherited is the ability of the organism to modify such a character."

BIOLOGICAL DIVERSITY AND ENDANGERED SPECIES

We may be able to discern evolution in process. But how do we observe extinction in process? Ordinarily, the answer is that we do not. For one thing, extinction is extremely infrequent over the history of life. For another, it is extremely difficult to predict with useful reliability. Finally, extinction is the last event in a long sequence of subtle ecological and evolutionary processes acting in concert.

Instead, we turn to the fossil record for clues into the processes of the past in expectation of bringing the lessons into the present and even taking them into the future. But "fossils open only a tiny obscured window on biological communities of the past."

Biological Diversity

Biological diversity is explained in terms of genetic variability. With greater specificity, the breadth of diversity based in profoundly simple chemistry is the result of sexual reproduction; structural alterations in karyotype; and chemical changes, mutation, within the DNA molecule. Traits are inherited and passed on to future generations, subject to selective pressures in the environment that Favour specific phenotypes over others. Survival and reproduction depend on the ability to adapt to those pressures.

Speciation or Extinction

The fossil record suggests that virtually all taxonomic lineages have become extinct without leaving any descendants in the form of new taxons. Extinction may well be the inevitable fate of every species. We have recently discovered that even cells extinguish themselves.

In fact, today's millions of living species are believed to have derived from only a small fraction of those living at any time in the distant past. Extant species represent less than 1%

of those alive some 600 million years ago. Actually, there are levels of what scientists identify as extinction.

Small, localized populations may die as a result of severe local conditions or the introduction of a particularly efficient predator. Such a local—ecological-extinction would not necessarily lead to extermination of a whole taxonomic lineage.

Larger local or full global extinctions also occur, especially in the face of competition from an ecologically similar species, because no two species can simultaneously exist indefinitely in the same niche. One species will be replaced, and community structure will undergo ecological or evolutionary adaptation. Even larger extinctions have occurred, presumably in a punctuating event such as that which wiped out the major taxon containing the dinosaurs.

THE ENDANGERED SPECIES ACT

In the year 1810, passenger pigeons numbered in the billions in this country. Just over a century later, they were extinct. Why? Because the passenger pigeon was known to gourmet and gourmand as squab. The human predator preyed on the population until it was eliminated from the ecosystem.

Some 200 years ago, taxonomists observed that certain extinctions had occurred in recent times at the hands of humans. Among these extinctions were fifty-three species of bird and seventy-seven of mammals. Exploitation was not the sole anthropogenic cause.

Some species were destroyed as pests; others were subjected to depredation by domestic animals or by organisms like the rat, which enjoys an embarrassing relationship with humans. Still other species were eliminated through the destruction of their habitat.

Chapter 11

Organisms and Environment Relation

POPULATIONS IN AQUATIC ENVIRONMENTS

A group of organisms occupying a given space at a particular moment in time is called a *population*. A population may consist of a single species and thus constitute a *species population*, or it may be composed of a number of species exhibiting similar ecological traits (food requirements, breeding sites, etc.) and regarded as a *mixed population*. In addition to structure and time-space attributes, a population also exhibits a number of measurable characteristics such as birth rate (natality), death rate (mortality), density, and capacity for increase. The extent to which these traits are manifested in an environment is strongly influenced by properties of that environment.

THE PRINCIPLE OF LIMITING FACTORS

The very fact that we can recognize an environment-organism complex such as a lake community or an estuarine community attests to the premise that all organisms do not react uniformly to all combinations of physical, chemical, and biological features of the environment. Through evolutionary processes, organisms have become variously adapted to certain sets of conditions.

It follows, therefore, that successful development and maintenance of a population depend upon harmonious ecological balance between environmental conditions and tolerance of the organisms to variations in one or more of these conditions. This idea suggests that one or more factors can serve to limit the areal distribution, density, and other attributes of a population. A factor that so checks or exerts some restraining influence upon a population through incompatibility with species requirements or tolerance is said to be a *limiting factor*.

You will note that the limiting factor principle rests essentially upon two basic concepts. One of these relates organism to environmental supply of materials needed for metabolism and growth. The second concept pertains to the tolerance which an organism exhibits toward environmental factors and conditions.

The aspect of environmental supply in the limiting factor principle includes, among other things, the provision of nutrients and other substances (respiratory gases, for example) necessary for growth of the organisms. If, out of a broad array of materials required for growth and development, one is absent, a given species will not be able to survive.

On the other hand, if the nutrient is present but only in small amounts, the population of a particular species will be limited proportionately. These conditions were described in 1840 by Liebig in what has come to be called the *law of the minimum*. This principle recognizes that the development of a population is essentially regulated by the substance occurring in minimal quantity relative to the requirement of the population.

It is frequently difficult to determine precisely which of the factors of a complex is exerting a limiting effect. This is due largely to the fact that the utilization of a certain substance may be regulated by other materials, or by physical factors in the environment, or by physiological adjustments of the organism itself. It has been shown, for example, that the uptake of phosphorus by the alga *Nitzschia closterium* is influenced by the quantity of nitrate and phosphate in the environment; nitrate utilization, on the other hand, appears to be unaffected by the phosphate. Potassium, calcium, and magnesium generally regulate interactions in plant metabolism.

The assimilation of nutrients by some algae is related to environmental temperature. Temperature and carbon dioxide tension may affect the rate of oxygen utilization in certain fishes. We have previously learned that the absence of free carbon dioxide in waters is not necessarily critical for certain plants which can assimilate carbon dioxide from bicarbonates.

Dissolved oxygen is necessary in respiration of most aquatic organisms. The minimal requirements of species vary and often limit the spatial distribution of certain forms in the community. The water issuing from the "boil" is oxygenless, but gains the gas with downstream flow. A small amount of hydrogen sulfide is present in the boil and decreases downstream; this may also be involved in the observed distribution.

The law of the minimum pertains primarily to the reaction of organisms to factors necessary for growth and metabolism. In the case of plants, these factors include the inorganic nutrients and solar energy for photosynthesis. In addition to satisfying their metabolism needs, organisms are also confronted with physical and chemical factors which essentially regulate the distribution and numbers and the extent to which the organisms are able to utilize growth factors. Such environmental features as salinity, temperature, and currents govern many populations.

The extent to which these factors limit a population depends primarily upon the *tolerance* of the organisms to a single factor, or, in some instances, to a complex of interacting factors. These ideas are the essence of the *law of tolerance* stated in 1913 by V. E. Shelford of the University of Illinois.

Tolerance to features of the environment varies widely among aquatic organisms. A population may exhibit a wide range of tolerance toward one condition, and a narrow range toward another. All stages in the life history of an organism do not necessarily show similar ranges of tolerance. The range of tolerance toward a given factor may be modified by another factor. A wide range of distribution of a species is usually the result of the wide tolerance.

In describing the tolerance of an organism, the prefix *eury* meaning wide, or *steno-*, meaning close or narrow, is added to a term for the particular feature. We have already become familiar with the terms *euryhaline* (wide salt tolerance) and *stenohaline* (narrow salt tolerance). To these, we might add *eurythermal* and *stenothermal*, pertaining to temperature. Let us now consider some examples of tolerance relationships, for such are of utmost importance in regulating the abundance and distribution of organisms.

Contractile vacuoles are characteristic of freshwater protozoans, but not of their marine relatives. In certain flatworms (*Gyratrix*, for example) the flamecell mechanism serving in osmoregulation and excretion is more complexly developed in freshwater forms than in estuarine ones; in marine flatworms, major parts of the system are absent. Freshwater crayfish possess long, well developed nephridial canals; in the estuarine and marine lobsters these structures are reduced and poorly developed.

The fluids of marine bony fishes are hypotonic to the environment, and therefore the animals tend to lose water to the medium. In contrast to freshwater inhabitants, marine fishes actively drink water, excrete a concentrated urine, and excrete excess salts (obtained by drinking sea water) against an environmental gradient through specialized cells in the gill region. In the estuarine environment, these marine forms must function in the manner of freshwater species.

The young of certain species of marine fishes inhabit estuaries and possess glomerular kidneys; with migration to higher salinity as adults, the glomeruli degenerate. Highly euryhaline fishes, particularly those adapted to extreme migrations from fresh to salt water (the eel (*Anguilla*), for example), are usually capable of performing as either marine or freshwater fish.

We have dwelt on this subject of osmoregulation because it demonstrates some of the forms of adjustment often necessary in order for organisms to tolerate certain controlling factors in the environment. In many instances, however, the limit of tolerance of a given organism may be modified by other factors.

It has been shown, for example, that the shrimp (*Crangon crangon*) and estuarine crab (*Carcinus maenas*) are able to tolerate lower salinities during summer seasons of warm water. In Japanese waters, the oyster (*Ostrea gigas*) exhibits a wider range of salinity tolerance at winter temperatures than during the warm season. The tendency for marine species to tolerate lower salinities at higher temperatures accounts, in part at least, for the generally greater number of species in estuaries of the tropical zones than in the cooler climes.

As in the case of salinity, temperature also acts as a controlling factor related to range of tolerance of species. In this sense, temperature serves to regulate growth and metabolic rates of various organisms, and often determines the time of reproduction. It is immediately apparent, therefore, that temperature is highly important in delimiting the rate of utilization of nutrients and light plants, and the tempo of food intake by animals, to satisfy metabolic demands. These relationships are contained in the "Q_{10} Rule," or Van't Hoff's principle, a chemical law applied to some physiological processes.

The principle holds that the rate at which biological processes proceed is increased nearly two-fold with each 10° rise in temperature (the rate is a linear function of temperature). Obviously, this principle operates only within the range of tolerance of a given species, and is further restricted by an optimal point within the over-all range. For example, if sufficient nutrients are available, an algal population may grow rapidly and "bloom" at the optimal temperature for that species; beyond this point, and before the maximum tolerance limit is reached, the population declines.

Throughout the living world, organisms have become adapted to temperature ranges and fluctuations in a spectacular variety of ways. In the first place, we recognize two broad divisions:

- The "warm-blooded" (*homoiothermic*) forms, the birds and mammals, whose body temperatures are maintained at a uniform level independent of environment; only a few of this group are wholly,

- The "cold-blooded" (*poikilothermic*) organisms whose body temperatures approximate that of the environment and vary accordingly. Among the latter group are such extremes as the algae that live in ice water of the polar regions and bacteria that inhabit hot springs at temperatures near 90°C. Between these extremes are found many species of plants and animals variously adapted to thermal conditions.

Great numbers of species are unable to exist as physiologically active stages throughout the range of temperatures experienced in their environment; these typically form spores, or other "resting stages," which exhibit a wider range of tolerance than do the active forms. Stenothermal and eurythermal species are found in all major taxonomic groups.

The European turbellarian, *Crenobia alpina,* is apparently restricted to a temperature regime below 12°C; sessile rotifers are usually found at temperatures above 15°C; tardigrades are active in a wide range from about 0° to 40°C; the optimum temperature for the bryozoan,

Pectinatella, is between 20° and 23°C, the lower limit of tolerance being near 15°C; the oyster (*Crassostrea virginica*) tolerates temperatures from 4° to 34°C. We quickly call to mind typical "cold-water" and "warm-water" fishes, many of these being restricted to such environments by the temperature tolerance of the developing eggs; the eggs of the eastern brook trout (*Salvelinus fontinalis*) generally fail to hatch at temperatures above 12°C, the optimum being near 8°C. The optimum temperature for growth of the adult trout is about 15°C. Thus, several optima may be operative in the life history of a given species.

Light appears to be a controlling factor in the distribution of certain animals in time and space. Diurnal vertical migrations of planktonic crustaceans have been noted and studied widely. These movements characteristically take the organisms to near-surface regions during the middle of the night, and to the depths during midday.

Numerous explanations have been suggested, and, indeed, the causes may be numerous. However, the correlation between the depth of penetration of the blue segment of the light spectrum and the position of greatest abundance of a copepod species, *Cyclops strenuus*, would indicate a strong influence of the light. Note particularly the conspicuous change in space occupancy with increasing day length from April to June.

From consideration of the "law of the minimum" as it pertains to nutrients, and the "law of tolerance" as it relates to environmental "controls" we can now appreciate the broader *principle of limiting factors*. This principle embodies the concept that population growth and success, in terms of abundance and distribution, are determined by a set of environmental factors, any one of which may, through scarcity *or overabundance*, be limiting.

We have seen a number of examples in which short supply of a requirement could hinder the success of a population; this is not difficult to understand. What is sometimes more difficult to comprehend is that an excess of normally important factors such as nutrient salts or sunlight can exert lethal effects upon organisms. For example, high concentrations of nitrate and phosphate have been shown to be toxic to both plants and animals.

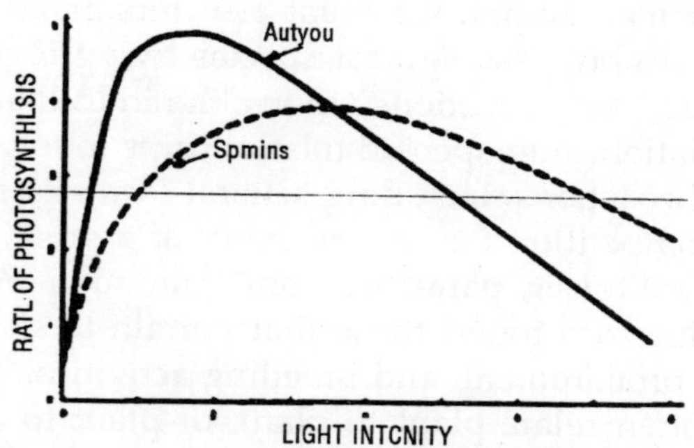

Fig. Photosynthesis (ìmoles CO_2 absorbed per 10 ìliters of plant matter per hour)

Several exciting aspects of the limiting factors principle can be shown in the reactions between sunlight and phytoplankton in photosynthesis. Note first that in both cases the rate of photosynthesis (measured by rate of uptake of carbon dioxide by plants) is inhibited by "too much" as well as by "too little" light. Light conditions optimums for photosynthesis are indicated at the peak of each curve, but notice that these conditions differ for the two seasonal assemblages.

It appears also that the spring phytoplankters exhibit a wider range of tolerance to the light intensities encountered than do those of autumn. Of particular interest is the higher rate of photosynthesis in the dim autumn light than in the relatively bright light of spring. This would seemingly indicate that the autumn community utilizes light more efficiently than do the spring populations. In all probability, however, the species comprising the communities are different.

The principle of limiting factors is not built upon physical and chemical factors alone. Biological features of the environment are often quite influential in determining the abundance and distribution, or over-all success, of a specific population. Once we introduce other organisms and their demands and activities into a set of factors, we immediately perceive greatly increased complexity of ecological relationships.

For in addition to reactions between a population and the physicochemical factors, we must also recognize coactions between, or among, the several species living together. We must also recognize coactions among the individuals of any given population. Interspecific relationships involve a broad spectrum of coactions, including several forms of symbiosis, tolerance, competition for one or many of a great variety of necessities, antibiosis, parasitism, and predation. Within this list of activities are found those that pertain to shelter, food supply and procurement, and breeding activities.

These often relate plant to plant, or plant to animal, or animal to animal in a broadly graded series of coactions ranging from those in which one member of the relationship suffers extreme harm at the expense of another, to those in which both of the participants are benefited. Singly, or in combination, interspecific relations may, and often do, act as a limiting factor. They are also basic in the structure and dynamics of communities. The ability of a species to successfully occupy a place in the economy of a community is also determined by intraspecific relations of the individuals and certain inherent biological qualities of the population. Intraspecific competition for such requirements as food, territory, and breeding partners may serve to limit a given population.

Similarly, overcrowding, which may in some cases lead to stress symptoms, can be limiting in nature. Inherent biological qualities such as frequency of breeding activity, number of offspring produced, survival rate, and mortality rate are often correlated with physiological tolerance to environmental factors. Upon these adaptations depend, at least in part, the abundance and distribution of a population. Some of these population attributes will be examined in the following paragraphs.

Although we have somewhat dissected the set of limiting factors for the sake of illustration, it is important to bear in mind that a single factor can become limiting only when it is near the maximum or minimum of the range of tolerance of a population. In nature, under usual conditions, a given population reacts to a complex of factors. The composition of this complex is not normally fixed; some of the components change seasonally, and a number of interdependencies exist, as we have seen.

A specific organism or population, on the other hand, presents a different story; its tolerance to each factor is more or less firmly established by inheritance and evolution. Thus the position and place of a species in the economy of a community is fairly well defined and delimited by relations between organisms and environmental factors. This thought brings us to another important concept in population ecology, that of the *niche*.

POPULATIONS AND THE ECOLOGICAL NICHE

We have just witnessed how an imposing series of environmental conditions and processes combine to circumscribe the total functions of a given animal or plant population. As a result of evolution of both the complex of environmental factors and the population, the functional reactions and coactions of a particular species have become essentially unique to that species.

Thus, each population (characterized by certain rather definite morphological and physiological parameters) and its environmental complex constitute a sort of system, the attributes of which distinguish it from other such systems. Once two systems overlap significantly, the population of each is thrown into competition for nutrient, or other, requirements, and one system breaks down. This "systems" example is a homely analogy to the functional concept of the *niche*.

Can more than one population occupy the same or overlapping niches in natural communities? This question has intrigued ecologists for many years. As yet, the problem has not been solved to the satisfaction of all. The idea is highly important in ecological thinking, however, because, among other things, it offers a point of entry into the study of the factors limiting a single population, and it also relates to events commonly observed in animal communities maintained in the laboratory.

In nature, these forms inhabit small ponds rich in organic matter, and the distributional ranges of the two species overlap. In laboratory cultures, D. *magna* consistently lost in the competition with D. *pulicaria*. Note that in both algae food and the yeast food culture, the population of D. *magna* reached its peak abundance at about 21 days, after which there was a rapid extinction due to interspecific competition. Observe also

the effects under the different nutrient conditions. In algae, the initial growth rate of D. *pulicaria* was quite rapid; in yeast, the rate was less rapid, being exceeded, as shown in the figure, by that of D. *magna*. Two limiting factors, food and oxygen, were operative in this experiment. The results clearly demonstrate that two species do not both persist in a simple environment exhibiting minimal niche diversification.

Under natural conditions, of course, food is but one of many factors for which competition by similar species may exist. Intense competition between closely related or ecologically similar species is often avoided by adaptations such as differing feeding times or breeding seasons, or by geographical distribution.

Another important aspect of the niche concept describes the "role" of a species population in the food and energy relationships in a community, i.e., as a "producer" or a "consumer." Although the dynamics and levels involved in these relations will be discussed more thoroughly in the next chapter, it seems worthwhile to mention briefly how these relate to our present topic. The role which a population performs in the economy of a community involves, again, a given set of reactions and coactions. The role of a diatom, for example, as a producer of the "original" energycontaining substance, is based upon the ability of the plant to carry on photosynthesis. The position, or niche, of the diatom, with respect to solar radiation, raw materials, and herbivorous organisms, is essentially the same throughout time.

Consider, on the other hand, that the role of a great number of animals differs greatly with various stages of life cycles. The important point here is that organisms may be rather narrowly and firmly established in a given niche, or, depending upon adaptations, may function in a number of niches during a life span.

ATTRIBUTES AND DEVELOPMENT OF POPULATIONS

In the foregoing discussions our interest has centered mainly on some of the major physicochemical and biological features of the environment and their effects on populations. How, in the face of such an imposing array of limiting factors, are abundance and integrity of a species population maintained; and, beyond this, how do numbers increase? The answer is found partly in environmental characteristics and partly in the unique attributes of the population.

The essential properties of the population which govern its growth, structure, and spatial distribution are *natality*, *mortality*, and *dispersal*. These qualities of the species are, in a sense, the "counterweights" set against limiting factors, the "degree of balance" militating, as it were, for or against species survival.

NATALITY AND MORTALITY

Natality is a population parameter which describes the rate at which new individuals are produced; it is, in essence, the birth rate. It is often important in certain population problems to consider two aspects of this attribute; these are:

- *Potential natality,* or the maximum rate of population increase if no limiting factors are operative,
- *Pealized natality,* or the rate experienced under natural conditions of reactions and coactions.

In estuaries, a population of the American oyster exhibits a very high potential natality, for a single individual is capable of spawning over 100 million eggs. The realized natality, however, amounts to only a small fraction of the potential. Estuarine populations of the sea catfish (*Galeichthys felis*) have a low potential natality; a single female seldom produces more than about 25 or 30 eggs at a time. Parental care, in which the male carries the eggs in his mouth until hatched (oral incubation), contributes to an apparent high realized natality for this species.

The value of the two aspects of natality is found in their usefulness in studying population growth and development in relation to environmental factors and the niche. Potential natality, determined from the rate and number of eggs produced per individual, serves as a standard with which a census of an existing population can be compared.

This measure alone, would not, of course, account for the momentary structure of a population. Natality is always a positive quality leading toward continual increase in population size. This effect is essentially offset by population traits such as mortality and dispersal.

Mortality, as a population attribute, pertains to the death rate of individuals of the population. In contrast to natality, mortality has a negative effect on population development, tending to bring about its decline. As in the case of natality, two aspects of mortality may be recognized; these are:

- Physiological longevity, or that experienced under the most favorable set of ecological conditions,
- The realized mortality, or that due to environmental effects prior to fulfillment of the inherited potential.

The realized mortality of a given population can be illustrated clearly by means of a "survivorship" curve derived from plotting the number, or percentage, of individuals still living at successive time intervals against the known life span. If a great majority of individuals of a particular population survived to the limit of the life span of the species, i.e., if physiological longevity were to be experienced, the curve would extend horizontally to physiological longevity and then drop steeply; such a condition is not found in nature.

DISPERSAL

Population dispersal refers to shifting or rearranging of individuals, small groups of individuals, or the entire population with respect to space and to time. Most species possess some adaptation for dispersal at some stage of a life cycle. In some forms, dispersal is accomplished during the seed or spore stage; in others the larval stage may be moved by active swimming or through passive transport by currents or other agents. Very few natural populations exist in which there is not some immigration and emigration of individuals. The degree of either of these may be quite subtle with respect to numbers and to the rate at which the movement occurs. In this case, dispersal would probably exert little effect on the over-all structure of the aggregation. Conversely, rapid emigration or immigration by large numbers may result in depletion or near-depletion, or in overcrowding of the original stock.

The effects of loss or gain of individuals are important considerations in management of natural populations, such as fishes, for human exploitation. For example, the removal of individuals by fishing of moderate intensity is usually followed by replenishment of the stock. The rate at which the population is rebuilt is, of course, greatly dependent upon the natality-mortality relationship. Extreme exploitation may reduce the numbers below a minimum level necessary for population regrowth. This condition can result in extinction of the population. Rapid immigration of large numbers of individuals often has deleterious effects on the original population, primarily through increased competition for breeding territory, cover, and food. For many years this idea was not appreciated in fishery management circles, and many bodies of water already at their maximum "carrying capacity" received truckloads of fish fry and/or fingerlings.

One result was overpopulation and stunting of all individuals of that particular species; another result was simply that the introduced fishes failed to survive or were eaten by carnivorous inhabitants of the lake or pond.

Under certain conditions natality alone may not be sufficient to maintain observed population numbers. In estuaries, for example, there is a net seaward flour of mixed water sufficient to compensate for the contribution of fresh water by the stream. Thus, estuarine plankton populations are subject to seaward flushing as determined by the rate of water circulation in the estuary.

In order to maintain a given population density, natality, mortality, and dispersal must be attuned to seaward transport. In Great Pond, an Atlantic estuary in Massachusetts, the summer population of copepods (Acartia) has been studied. It was found that the reproduction rates in the uppermost parts of the pond were not adequate to account for the numbers of animals there, that immigration from deeper layers contributed to the population density of the upper region. In the middle portion of the estuary, natality was apparently sufficient to replace the loss by seaward transport. At the lower end of the estuary, great mortality was experienced, the small population present there being maintained only by immigration of copepods from the upper regions.

As indicated earlier, dispersal, natality, and mortality together constitute a set of interacting factors which essentially regulate growth and structural dimensions of a population. We have seen that there is a tendency for populations to grow, due to the fact that the potential natality normally exceeds the potential mortality.

However, the extent to which such growth proceeds would be tempered by dispersal traits. Should the realized natality exceed the realized mortality under conditions of minimum dispersal activities, the population would increase. Under the same natality-mortality conditions, and with excessive emigration, the population could be expected to decrease.

With equal birth and death rates, the population would tend toward stability, but again, this is influenced by movement of individuals. If the realized mortality exceeds the realized natality, extinction of that population would be expected; recruitment from nearby aggregations of the same species could serve to maintain the "unbalanced" population. Having given attention to these "forces" which mold population growth and density, let us now turn to consideration of these two population parameters.

GROWTH OF POPULATIONS

All natural species populations possess an *innate capacity for increase* at a maximum rate—the *biotic potential*. For a given species, this maximum rate is set by inherited potential natality-potential mortality values. These values differ greatly with species. Compare, for example, the figure given previously for a single oyster, with that of man. In nature, the biotic potential is not normally fulfilled, due to the limiting features of many environmental factors.

Thus, opposing the innate tendency of a population to increase is a set of conditions which, in effect, offer resistance to population growth. We recognize, therefore, that the growth and success of a species population depend upon the degree of "harmony" struck between biotic potential and *environmental resistance*.

The actual, or realized, rate of growth of a population is determined by the initial number of individuals, plus additions, minus the number lost per unit time. If, during a period of one year, two individuals produce 24 offspring from which four are lost, the population density would stand at 22 at the beginning of the following year. If similar conditions of natality and mortality prevail during the second year, the census should be 242 individuals. In other words, the rate of population growth is geometric, and is represented by the relationship:

$$\frac{dN}{dt} = rN$$

in which N is number of individuals and t is time.

The symbol r represents the rate of natural increase of a population, in other words, the difference between natality and mortality at a given moment;

$$\frac{dN}{dt}$$

Gives the average rate of change of N per unit time. If such growth continues unlimited, it takes the form indicated by the curve in Figure.

Since a particular community can support only a given number in any one species population, it is apparent that the maximum biotic potential can not be maintained in nature. Consequently, the form of the growth curve of populations subject to limiting factors must differ from the "unlimited" curve in Figure, and so it does. As a population begins to increase under favorable environmental conditions, growth is typically rapid and passes through a *phase of accelerating growth*.

This is especially characteristic of populations exhibiting high natality and under conditions of little intraspecific competition. In time, the rapidly increasing numbers do create problems of maintenance, and the growth rate begins to slacken, the population passing through a phase of *decelerating growth*. The point on the curve at which the rate of increase slackens is called the *inflection point*.

Eventually the population size becomes somewhat stabilized through essential equilibrium of reactions and coactions.

This size level may be at, or near, the *carrying capacity* of the community. On the curve, the carrying capacity (number of individuals) is designated the *asymptote*. Note that the curve representing population growth under natural conditions is typically in the form of a shallow "S," or the *logistic curve.* Given selfregulating growth processes, the curve may be derived from the differential equation:

$$\frac{dN}{dt} = rN\frac{(K-N)}{K} \; or \; \frac{dN}{dT} = rN\left(1-\frac{N}{K}\right)$$

Where τ is the biotic potential for each individual in the population; N is the total momentary population size; t is time; and K is the maximum number of individuals possible in the population under prevailing conditions.

$$\frac{(K-N)}{K}$$

The expression describes the extent of resource utilization by increasing population; in other words, the growth of the population itself causes the environment to become more limiting. As a result of this condition, the potential rate of reproduction is decreased as the population size grows toward the asymptote. This equation tells us that the population growth rate is equal to the maximum possible rate of increase times the degree of realization of that maximum potential rate.

The equilibrium which a population maintains about the asymptote is primarily a function of natality (*n*) and mortality (*m*), and, under certain conditions, dispersal. At equilibrium, n-m = 0, meaning that losses due to mortality are offset by additions derived from births. Deviations above the asymptote occur when *n-m* is positive (population increase); deviations below the asymptote follow when *n-m* is negative (population decrease).

In the growth of a laboratory population of a small fish, the guppy (*Lebistes reticulatus*), are fitted to the logistic curve. Note the characteristic phases of growth until the asymptote is attained. Observe also that the population increase initially "overshoots" the asymptote, but returns to the carrying capacity; this phenomenon is often encountered in population studies.

Upon reaching its asymptote, a population does not normally remain at uniform density, but rather fluctuates in numbers in response to one or more ecological or inherited factors. Irregular fluctuations, or relatively asymmetrical departures from equilibrium, may result from variations in any critically limiting environmental factor. A great number of observations on populations of plants and animals have shown symmetrical fluctuations, or oscillations, of population numbers in response to factors regulated by daily rhythms, tidal cycles, seasonal and annual cycles, and to predator-prey and host-parasite relationships.

Within a given population, the relative number of individuals in each age level is important in determining the growth form, as well as serving as a basis of diagnosis of the condition of the population. These general diagnoses assume that the distribution of age groups in the population is stable. The importance of age distribution in a population lies in the cumulative effect of the natality-mortality values for each of the age classes on the general population.

As indicated above, stability of population structure depends upon nearly equal rates of made reference to an oligochaete population with a density of 19,000 individuals per square meter of bottom surface in the saline region of an estuary. Population density is also time-relative in that the number, or mass, of individuals may, and often does, vary temporally. Daily vertical movements of certain zooplankton, for example, result in changes in population density at a given level in a lake; seasonal migrations of some fishes and other animal cause fluctuations in population size. Certainly, marked changes in population density are observed during breeding times.

An observed population size represents the product of a number of environmental features and relationships working through biological processes. Thus, analysis and study of density yield mush enlightening information on growth form and the effects of specific environmental factors on population development. It is often difficult, however, to recognize and measure subtle effects operative under natural conditions. Ecologists frequently surmount such difficulties by laboratory studies under controlled conditions, and by the judicious use of statistics to test the meaning and validity of laboratory and field observations.

Although influenced to a considerable extent by biological attributes of the species and the biogeochemical features of the environment, population density, per se, exerts effects on the members of the population and upon the environment. The density of a given population in a prescribed space may affect the population rate of increase, the proportion of males and females, the growth rate of individuals, the rate of utilization of nutrient substances and respiratory gases, and other processes.

Thus, there is an optimal density level for the growth or success of a population, and this implies that there must also be proportional limitation of growth or success under overcrowded or undercrowded conditions. It has long been known that there are deleterious effects on growth, reproduction, survival, and other biological processes which accompany overcrowded situations.

It remained for W. C. Allee and his students to demonstrate that at all levels of the animal kingdom there is added safety in numbers up to the optimal population level; there are adverse effects accruing to undercrowded populations. This is known as the *Allee Effect*. One example of this obtained by noting, the relative rate of population increase. If data from a logistic population curve are replotted to show the change in rate of increase (dN / dt) against density, one can see that at the low and high densities there is a low rate of increase while the highest rate of increase occurs at some optimal population level.

There may be several mechanisms whereby aggregations of individuals can grow and survive better than isolated individuals in the same medium. One such mechanism involves decrease of exposed surface. In the group, there is less surface area per individual exposed to the environment than would be if the individuals were dispersed. For example, the individuals in a group of flatworms have less total surface exposed to adverse effects of ultraviolet light or other extreme environmental factors than would be the case if these individuals were dispersed and exposed.

Similarly, masses of sea-urchin sperm survive loner than

dispersed cells since there is less environmental exposure to the sperm inside the mass. It might be noted, also, that there are undercrowding effects on the embryonic development of sea-urchin eggs.

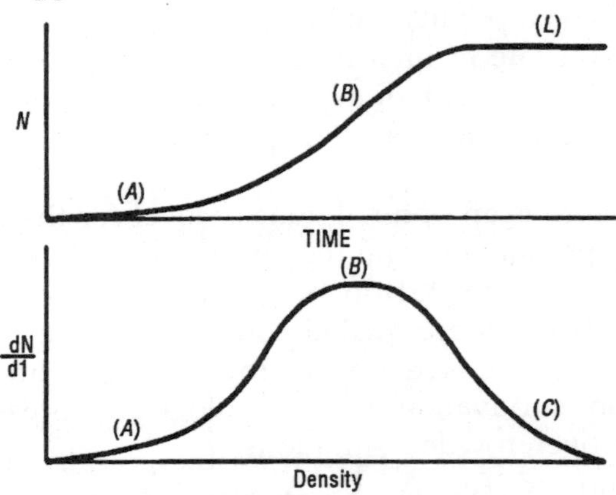

Fig. Schematic Representation of the Allee Effect, in which data from the logistic curve (above) are plotted to depict rate of population change (*dN / dt*) as a function of density (below).

Another aspect in the success of groups versus individuals is that the group can more efficiently condition the medium by removing poisons, changing the salinity of the water, or by adding growth-promoting factors. It has been demonstrated that groups of goldfish can survive longer than isolated individuals when placed in media poisoned by silver nitrate; the group more quickly removes and detoxifies the poisons, whereas individuals usually die. The groups of goldfish also grow faster as a result of sharing particulate food that spews from their mouths, and as a result of a higher concentration of a growth factor associated with the slime produced by the fish. Other studies have demonstrated that groups of protozoa have a higher per-individual growth rate than do isolated individuals due to the greater production of some growth-promoting factor, and also because the group can better control the population of bacteria.

There may also be benefits to the group resulting from more efficient coordination of activities that produce greater synchrony in reproduction. This heightened psychological state that leads to greater sychronization of activities and consequently to less mortality, has been shown for shore-bird colonies.

DISTRIBUTION OF INDIVIDUALS OF A POPULATION

We have previously considered dispersal of population individuals as a process and condition contributing to population growth form. Of utmost importance to population dynamics generally, is the manner, or pattern, in which members of a population are distributed within the community. This quality of populations is also of primary concern to the study of intrapopulation relationships.

Within the space occupied by a given population the individuals may become dispersed in a *random* pattern, in an *evenly-spaced* pattern, or as *aggregations*. Obviously, the pattern exhibited by a given population is subject to considerable variation depending upon the census of the population, the number of sites available for occupancy, food accessibility, and a host of other factors. Equally important in setting the dispersal pattern is the inherent sociological nature of the species at various seasons. Important as the problem of animal distribution is, it has received insufficient attention, especially as pertains to aquatic organisms. Future researches may well reveal additional patterns of intrapopulation distribution.

Completely random distribution of individuals is apparently uncommon in nature. The degree of "randomness" in the dispersal of individuals is determined statistically by several methods.

One of these involves the fitting of observed distributional data to the "Poisson series," a statistical measure of the frequency with which aggregations of 0, 1, 2, 3, 4, 5,... *n* can be expected in plots or quadrats. Deviation from the Poisson indicates various patterns of nonrandom distribution. It has been suggested that the lack of random dispersals in nature results from a generally inherited tendency of most animals to aggregate to some degree.

Evenly spaced distribution patterns are probably best explained on the basis of tendencies of many species toward *territoriality*. As an animal trait, territoriality refers to Defence of a given area by overt aggressive action. The rather euryhaline cyprinodont fish, *Cyprinodon variegatus*, exhibits marked territoriality during breeding season. At this time, the male fish come to occupy vigorously defended areas in the shallow zone of marsh streams. The territories are small, usually being about 30 to 50 cm in diameter, and the aggressive Behaviour lasts for several days; any male intruder is immediately attacked by a territory holder.

The females are usually somewhat concentrated a short distance from the area of territories, and show little excitement. With considerable frequency females singly enter the territory of a male, where a brief breeding encounter takes place. Territoriality is often associated with dominance-subordinance levels developed in many animals. Green sunfish (*Lepomis cyanellus*) in a laboratory colony develop hierarchies and defend their respective positions. Depending upon numbers of individuals and available territories, the most subordinate fish may not be able to establish a place in a "peckorder" such as do members of chicken flocks.

The *home range* of an animal is the region in which the individual normally travels. The important distinction between territoriality and home range lies in the aggressive response to invasion of the area shown in territoriality; home range carries no connotation of aggressive Behaviour. The concept of home range has been investigated less in aquatic animals than in terrestrial forms. However, a fair volume of literature on the movement of tagged fishes in lakes and streams strongly points to homing and home range in these animals.

For example, of 4557 fishes marked and released at Grapevine Point, Douglas Lake, Michigan, from 1937 to 1939, not a single one was caught from any other point. Investigations in an Indiana stream revealed little movement of fishes between pools. Most recently, results from a study of homing in the newt (*Taricha* (= *Triturus*) *rivularis*) in a California stream convincingly attest to observance of home range in this semiaquatic species. Even blinded individuals returned to their original locality after having been displaced some 0.8 km downstream.

The tendency to form aggregations is widespread in both plants and animals, and results in the most frequently encountered pattern of distribution. These aggregations, however, may be somewhat randomly scattered over a given space. Aggregation has been described as "a uniform response to a nonuniform environment"; it may also be a response to inherited social traits. One can call to mind a wide array of aggregation patterns, ranging from pronounced "schools" of certain fishes to tightly formed clumps of flatworms on the undersurface of a stone on a stream floor.

Time and space do not permit us to give due consideration to the many aggregation patterns. As to causes of this general type of distribution, we can list: annual, seasonal, or nocturnal-diurnal variations in physicochemical qualities of the environment; coactions among biological components of the community; physiological processes such as reproduction; and the above-mentioned sociological attributes of the species. The effects of aggregative tendencies are varied and manifold; some of these have been presented in the preceding discussion of population density.

THE SPECIES, UNIQUELY UNIQUE

Throughout history and in a variety of contexts, the question has come up: Just what is the human species? A notoriously superficial answer to what a human is relates to the elements of which the human body is composed.

Like all organisms, we are composed mostly of carbon, nitrogen, hydrogen, and oxygen, with the last element constituting nearly two thirds of our total weight. In fact, if sodium, calcium, phosphorus, potassium, chlorine, sulfur, and magnesium are added, 99.9% of our constituents have been taken into account. An array of heavy metals from aluminum to zinc represents the remaining minute percentage. This litany of chemicals, however, is not much of a start on any definition of this species. By definition, every species is unique, but is there something about *Homo sapienssapiens* that actually makes this species *uniquely* unique among species? Many of the contexts in which this question arises are well outside the scope of a quasi-scientific ecology text. But two contexts are particularly relevant: the scientific and the political.

Ecologically, the human species occurs as one or many populations distributed throughout the globe and occupying the most flexible niche of any species. Politically, the human species is perceived as a particularly devastating organism potentially capable of destroying whole ecosystems and perhaps the biosphere itself.

We may be unique in our ability to adapt to an extraordinary range of environmental conditions, not through physiological or other "purely" biological means but through intelligence and our ability to communicate abstract ideas, including concepts of the past and future. We can learn from the past of our species as well as from personal experience. We can plan for the future, often by applying past history. Language, writing, culture, and both an individual and a political sense of morality may be features that render our species quantitatively and qualitatively unique in a sense no other species is unique. Our minds represent a new level of organization as profound a leap from life as life is from physicochemistry.

Human horizons are constantly expanding, literally scaling new heights and exploring new dimensions. We see ourselves as Lilliputians, poised precariously on a very small planet. But through intellect, we "have the extraordinary ability to examine and comprehend the structure of the universe. This [alone] truly distinguishes us from less-evolved animals." The potency of the human mind is one lesson from contemporary astronomy.

With this recognition combined with concepts of the Gaia Hypothesis or of deep ecology comes a paradox for all of us but most particularly for environmental professionals. If our species is a part of nature, our actions cannot be other than natural, no matter how detrimental to our ecosystems. If our cultural organization distinguishes us from other life forms, how can we be bound exclusively to follow "nature's way"?

In the end, it would seem that evolution has led to human nature, which has led to culture; culture has enabled the Behaviour that is believed by many to impair nature's integrity. We face a quandary in that any return to nature for us would contravene our nature and our history. In seeking a way out of our quandary, it is worth contemplating the ecologically sound assertion placing populations beyond good and evil. It is the nature of every population to grow to fill its ecosystem. When too numerous, organisms either die out or transcend themselves.

Those that transcend "find new ways to procure room, carbon, energy, and water." But they produce new wastes, which test them. At what stage is the human population? *Are* we too numerous? If so, are we at the brink of extinction? Or are we about to transcend ourselves?

THE POPULATION ECOLOGY OF THE HUMAN SPECIES

At least one ecologist holds the human species unique in our capacity to alter our niche seemingly at will. We can increase the sources of niches in deriving energy from something other than complex organics we ingest, and we can use novel materials and systems unlike any other species. Another scientist finds the combination of size of brain, social structure, speech, and the use of tools to be what sets humans apart.

In other words, the biological system we call culture is unique to us and a new level of organization. It is culture that allows us to manipulate our environment in ways that are both purposeful and anticipatory. It is in culture that our unique capacity for abstract reasoning—from mathematics to aesthetics-comes to its culmination. The strategy of human cultures, each a suprapopulation, is to improve the quality of life for one or more future generations.

Sometimes our biological nature and our cultural one are in conflict. Biologically, it was expedient when some thousands of years ago we attained the capacity for which every species is striving: the ability to alter the environment to eliminate other species. Morally, we are beginning to take a second look. Doing so does not come easily. Has Garrett Hardin's "tragedy of the commons" placed us in need of "lifeboat ethics"? Must some individuals be denied a place on the global lifeboat for the sake of those already on board?

Indeed, we may not yet have escaped our first niche. We started out as wandering ice age foragers, and our physiques, temperaments, desires, and patterns still reveal that beginning. There are those who consider humans either an unnatural population or, perhaps more accurately, a supranatural population. There are others who perceive urban human aggregations as akin to a climax community. Whatever the scientific reality and whatever the political, economic, and social ramifications of those realities, *Homo sapiens sapiens* is a particularly important biological system to environmental professionals.

If one is to analyze humanity in ecological terms, one of the first questions to arise is whether the species is subject to limiting factors. Or does that seemingly infinite capacity to adapt into expanding ranges of tolerance, through seemingly infinite expansions of consumption, have a maximum? Maximum or not, can human populations thrive under near-optimum conditions without compromising either the carrying capacity of their respective ecosystems and ultimately the biosphere itself? Are we really jeopardizing other species in some "unnatural" sense, or are we merely manifesting the traits of all species in filling and expanding our niche?

Significantly enough, we may be the only species capable of asking such questions and moved by a sense of obligation associated with the answers. As Odum has observed, we must *transcend* the principles of general ecology because of our unique abilities to control our immediate surroundings, and our tendency to develop culture independently of environment.

First, we must recognize that human populations are parts of widely ranging biotic communities and ecosystems.

Superimposed on that status is the reality that our ecology goes beyond mere demographics—the statistical functions of any population. Causes and effects driving our populations demand closer analysis than statistics alone can provide. Finally, internal dynamics as well as relationships to external factors must be taken into account. Where biological regulation may not succeed, cultural regulation may be demanded.

We are responsible for any number of perturbations in ecosystems. Among the most frequently recognized are nutrient enrichment, acidification, toxification, nutrient depletion, habitat modification and outright destruction, and both the elimination of native species and the introduction of alien species. Above and beyond assumption of our niches, our intrusions into ecosystems can be stimulatory or inhibitory or can impose structural alteration.

In considering the ecology of human populations for purposes of application, most particularly in the marketplaces and in law and policy, it is essential to expand the questions and any answers derived beyond ecology into the cultural fields that do make us "uniquely unique."

When we perceive anthropogenic ecological perturbation, the first step is to comprehend the ecological results of that perturbation. But before we can turn to cultural institutions to prevent consequent harms or to protect or restore the ecosystem, we need to explore the cultural implications of the perturbation, of its ecological results, and of available responses. Only then can we begin to take effective action.

When we do act, we must always remain sensitive to ecological complexity. We must heed the lessons of past intrusions into ecosystems, whether taking the form of intervention or representing a perturbation. The former is presumably, but not necessarily, beneficial to the system, the latter detrimental. The truth is, we seldom are able to predict whether an action will be intervention, intrusion, or perturbation.

Too often, whatever the intent or lack of it, our actions result in ecological backlashes. These are unexpected and certainly unintended ecological responses. In simple terms, ecological backlashes can be categorized as failures to secure the intended correction or as additional perturbation in the subject ecosystem or one somehow connected to it.

In short, environmental professionals and, indeed, the population at large must heed all the ecological principles and concepts described here. We must take them personally! Only then are we in a position to apply our cultural selves to resolving our place in the biosphere.

SYSTEMS ECOLOGY AND RESOURCES

In seeking to apply ecological principles to human activities, some rhetorical questions should lie behind those efforts. What are the practical implications of the phenomenon that the optimum for quality is always less than the maximum population that an ecosystem can sustain? To what extent are deteriorating environments attributable to cultural tendencies to be too independent of the natural environment?

Systems ecology, emergent with the analytical power provided by computer technology, is a kind of return to the black box described in the opening chapters. When attempting to analyze the ecosystem or some component thereof, the ecologist or the environmental professional is immediately confounded by the sheer complexity of the interactions at and among all the relevant levels of organization.

As should be thoroughly instilled by now, every level of organization from pure mathematics to planetary ecology and beyond has something to contribute to an ecosystem's output as a result of each and every isolated input. In a very real sense, systems ecology and computer modeling are founded in the Theory of Integrative Levels.

Systems analysis is defined for purposes of ecological systems as the process of translating internal physical or biological concepts into a set of mathematical relationships that are then manipulated to predict or explain the Behaviour of that system.

Some of the complexity can be transformed into the mathematical symbols and equations on which computer modeling depends brief, the mathematical symbols can be a useful shorthand describing complex ecological systems. The equations can permit a formal statement (which can often be transformed into a highly effective visual presentation) of how an ecosystem's components are likely to react to change.

The model resulting from the mathematics derived through systems analysis remains a mathematical system. However valuable, it will always present an imperfect and abstract simplification of the real world.

The challenge, of course, is to work with that relatively simplistic representation while recognizing its imperfections in order to serve the ecosystems upon which we depend for life. As every research scientist can avow, even tentative answers open the way to a fuller understanding of how the natural world functions, allowing for predictions and some degree of control.

But it is the real world that is correct, if there is any conflict between the answers and the underlying reality reflected in the information provided by the natural sciences. Wrong answers and other "failures" serve to reveal flaws in the conceptual framework and may point the way to corrections in the modeling process. If applied to a living system, however, wrong answers can do more harm than good, intentions notwithstanding.

The criteria to be applied in evaluating models are threefold: realism, precision, and generality. Realism is the degree of correlation with the biological concepts represented by the model. Precision is expressed through the ability to predict quantitative change and to mimic the data upon which the model is based.

Generality is the breadth of applications the model fits. Models are also characterized in terms of resolution and wholeness. The former expresses the number of attributes included in the modeling process; the latter, the interactions recognized between levels of organization.

Allegedly, ecological models are noted more for their generality than for their precision. There is an argument to the effect that a model cannot at once be general and precise. Any improvement in the one is matched by a loss in the other.

Moreover, an ecological model should be expected to be unrealistic in terms of lower levels of organization because of the lack of complete information concerning input and output at those levels, the sum of which are ultimately being expressed at the level of interest.

Whatever the tools of the environmental professionals, it is important to remind ourselves constantly to expand our thinking and to take care in imposing restrictions in that technology alone is not enough. "[M]oral, economic, and legal constraints arising from full and complete public awareness that man and the landscape are a whole must also become effective." Environmental professional or private citizen or both, it is essential further to heed the ecological principle that the optimum for quality for a population is always less than the maximum quantity that can be sustained.

Another ecological principle that must be heeded is that the more we demand of an ecosystem, large or small, the less energy is retained by that system for maintenance. Has *Homo sapiens sapiens* been justifiably labeled a parasite on the environment at large? Are we truly as destructive to our host as most parasites?

HUMAN NATURE ENVIRONMENTAL PHILOSOPHIES

Throughout the history of the human species, environmental philosophy has played a major role. That role has, however, undergone profound alteration, especially in Western civilization as it is manifesting itself and merging with others here in the United States.

We have gone from huntergatherer to tiller of the soil to industrialist to scientist to the age of information. Both our niche and our intellectual relationship to our ecosystems have altered accordingly. Emerging from the status of one more primate species in the wild, we have learned to fear the natural world. It is in our nature to seek conquest of whatever we fear.

Some may believe human populations have conquered the natural world, but most of us realize that our place as conqueror is tenuous at best. In recent years, we have come to realize that we may represent a unique hazard not only to our ecosystems but to the biosphere as a whole. We are rediscovering our niche and the habitats in which it is active.

When humans left the forests and fields to enter into agricultural and later societies, they came to fear the wilderness. When North America was invaded by representatives of Western civilization, the frank intent was to conquer those usurping the land—wild beasts, howling wilderness, and hellish fiends. This vast wilderness was a place of evil as well as one of physical danger. Conquest was a moral imperative. The contemporary environmental movement in the United States traces its most recent roots to the conservation and preservation movements in the early twentieth century. In fact, early conservation helped shape the notion of "commons" as a public domain in resources that are held not susceptible to private appropriation.

Conservation seeks to ensure the preservation of a quality environment in company with a continuous yield of useful plant, animal, and natural materials—natural resources. Conservation, then, tends to be more anthropocentric than preservation, which tends more toward biocentrism or ecocentrism. Our literature and the popular arts in the United States are rich in images identifiable as personal ideas of wilderness, conservation, and preservation. Among the best known of the writers is Aldo Leopold, who spoke of conservation as a "state of harmony between men and land"—his "land ethic," an ecological conscience.

Ecological health, according to Leopold, lies in the land's capacity for self-renewal. Conservation arises from efforts to understand that capacity in order to preserve it. The land ethic is a major concept incorporated into the contemporary environmental movement.

THE CONTEMPORARY ENVIRONMENTAL MOVEMENT

What Is It? In the early 1960s, the environmental movement in the United States still took the form of conservation or preservation. In the 1970s, the shock of gross pollution and disappearing landscapes moved us to a kind of grim pleasure expressed in part as cultural and individual ambivalence toward progress. The environmental movement provided one outlet for our malaise.

Whatever our individual relationship to the movement, we can all probably agree that the role of culture is not to overcome nature but to be open to its potential. A primary lesson from ecology is that all "poverty" is attributable to unrestrained population growth. The environmental movement, then, is seeking ways to alleviate the results of population growth, unrestrained at least at the local level, and thus to relieve ecological poverty in whatever biological systems in which it is being expressed.

An Ecological Perspective

To the extent Odum can be cited for the ecological perspective, he has observed that ecologically sound conservation, one of the missions of contemporary environmental policy, is neither a matter of hoarding resources nor one of total rejection of development. Instead, the objective should be one of rejection of unplanned development in violation of ecological laws.

In the past, such ecologically unsound development has rightfully been attributed to ignorance as well as to shortsightedness and advancement of parochial interests.

Today, there is no excuse for development characterized by any of those three factors. In the United States, at least, there is a very real movement toward integration of all the factors supporting development that is ecologically responsible. It is a youthful movement, however, and subject to all the excesses of youth. At least we are moving away from all three impediments to environmental quality.

One of the major problems, and one that may be insurmountable, is the nearly total focus on the human perspective. We tend to speak in terms of environmental quality in association with products, recreation, and aesthetics. We seek to ensure continuous yields of useful plants, animals, and materials. Not until we act on the knowledge that ultimately it is only healthy environments that are healthful are we likely to change our ways in an ecologically sound manner. As environmental professionals seek to serve what is sometimes called stewardship of ecosystems and of the biosphere itself, some turn to cultures not ordinarily included in that of mainstream America. For example, some assert that we should heed the injunction of certain Native Americans to consider the effects of our actions through seven generations. All too often, we have barely thought beyond the current generation.

Another lesson to take from Native American peoples is the notion that we cannot preserve ecological systems by the sole expedient of dictating by law. There instead must be a profound alteration in the way each of us relates to the world around us.

Chapter 12

Nature of Water

It is truistic but nonetheless meaningful to say that the most important single feature of lakes and streams is water. Water is a biological necessity of plants and animals. It is required by plants for photosynthesis and by both plants and animals in general metabolism. Water in the lake basin or stream channel is of basic import ecologically, for it is the medium in which the members of the community live.

This is to say that the inhabitants must be adapted for finding shelter and food, utilizing dissolved gases, and reproducing in an environment very different from the terrestrial one. As an environmental medium, water enters into and maintains the total economy of the ecosystem.

Therefore, before considering the physical and chemical features of inland waters and estuaries it seems proper to give brief attention to some of the properties of water as such. These attributes are basic to the environmental processes and characteristics of waters in lakes, streams, and estuaries.

THE WATER MOLECULE

The water molecule is constructed from covalent bonding of two atoms of hydrogen (H) and one atom of oxygen (O). In combination these give the familiar H_2O. Two chemical bonds unite the two H atoms to one O atom in the form H-O-H. The nucleus of each atom is positively charged; negative electrons revolve, or orbit, around the central core.

The path of an electron is called an *orbital,* or energy level, and the bonds between a hydrogen and oxygen atom are formed by the overlapping of the orbitals in a fashion such that each atom contributes one electron which, in effect, completes the orbital of both atoms.

Four of the unshared electrons of the oxygen atom, important in some reactions, are located considerably farther from the nucleus than are the remaining two. Because of the greater positive charge on the oxygen nucleus, the bonding electrons are not distributed equally between the hydrogen and oxygen nuclei, but are relatively closer to the oxygen atom. This results in a partial positive charge on the hydrogen atoms and a partial negative charge on the oxygen atom, or polarity.

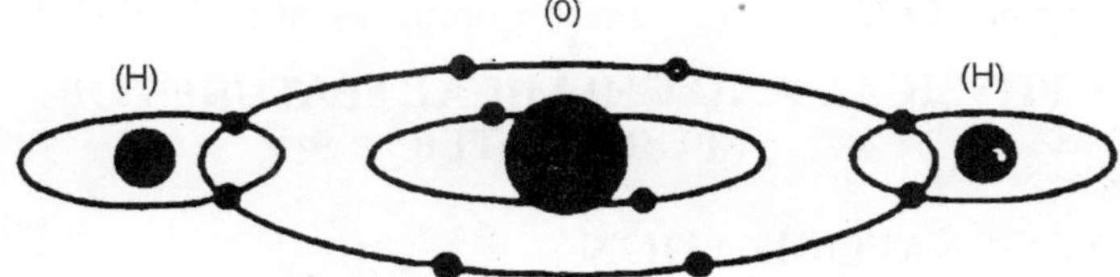

Fig. Diagram of the Water Molecule Showing the Overlapping Orbitals and the Shared Electrons

Water is a highly organized liquid. The molecules are arranged in a definite configuration. The association of the molecules by hydrogen bonds into chains giving liquid water or ice, and the dissociation of the molecules to form Vapour, account, in large part, for the versatility of water. Chemical reactions such as oxidation and hydrolysis occur when chemical bonds between the hydrogen and oxygen are broken. Physical processes, such as the melting of ice and evaporation, take place when the attractive forces between water molecules are broken without affecting the molecule.

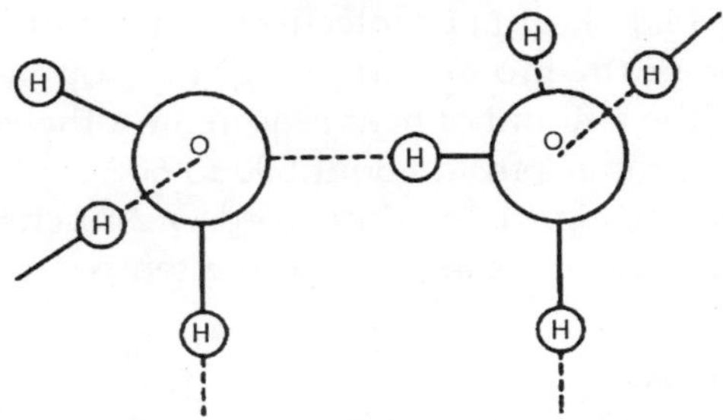

Fig. Two Molecules of Water Joined by Hydrogen Bonding

Chemical bonds uniting two hydrogen atoms with one oxygen atom are shown as solid lines. Hydrogen bonds are represented by dashed lines. This is but a portion of a vast system of variously oriented molecules in liquid water.

Ordinary water is not merely an aggregation of individual water molecules composed of two hydrogens of atomic weight 1, and an oxygen of atomic weight 16. Water also contains isotopic substances such as "heavy water" in which two deuterium (H^2) atoms are combined with an oxygen. There are also lesser amounts of isotopic molecules in which O^{17} and O^{18} replace O^{16}, and very small quantities of tritium (H^3).

PHYSICAL AND CHEMICAL FEATURES OF PURE WATER

HEAT OF VAPORIZATION

Compared with other liquid compounds of similar, simple molecular composition, water vaporizes very slowly when heated, i.e., water has a very high latent heat of vaporization (evaporation). Vaporization is a process in which thermal energy is supplied to overcome the attractive forces between water molecules (mainly hydrogen bonds in liquid water). A given volume of liquid water contains a larger number of hydrogen bonds per unit volume than most other common solutions.

We have just seen that there are two hydrogen bonds linking each water molecule to its neighbors; therefore the heat required for evaporation is approximately twice the hydrogen bond energy (4.85 kcal per molecular number of H bonds, or 9.7 kcal per gram molecular weight). Depending upon temperature, the amount of heat required for the vaporization of 1 g of water ranges from about 500 to 600 cal (9700 cal per gram molecular weight). In other words, as much heat is used to vaporize 1 g of water as to raise the temperature of 540 g by 1°.

SPECIFIC HEAT

The specific heat of a substance refers to its capacity to

absorb thermal energy in relation to temperature change at constant volume. Water holds a great amount of heat with a relatively small change in temperature. The unit of measure of specific heat is the gram-calorie, or small calorie, which is defined as the amount of heat required to raise the temperature of 1 g of water from 14.5° to 15.5°C. Thus, water is the basis of this parameter and is said to have a specific heat of 1; the specific heat of other compounds is calculated as the ratio of their heat capacity to that of water. Only a few substances, including lithium at high temperatures and ammonia, have a higher specific heat than that of water.

This heat capacity of water has far-reaching implications. For one thing, it acts as a buffer against wide fluctuations in temperatures, thereby ameliorating terrestrial climates near large bodies of water. Secondly, an aquatic organism is subjected to much narrower ranges of temperatures than a land form: whereas land areas may reach 38°C or more, lakes seldom exceed 27°C. Furthermore, the slow rate of seasonal cooling and warming of lakes and streams is attributable to the high specific heat of water. For the same reason, changes in lake temperatures lag behind atmospheric fluctuations. In natural waters, dissolved substances lower the specific heat.

Much heat is necessary to bring water to boiling because a considerable amount of heat is dissipated in energy of vaporization before the boiling point is reached. The "reluctance" of water molecules to separate and emerge as Vapour is due, as we have seen, to the hydrogen bonding of the molecules. Methane (CH_4) has nearly the same molecular weight (16) as water (18) but boils at-161°C as compared with 100°C for water. The difference is found in the fact that the outer electrons of the methane molecule are bonded chemically, there being no free pairs of electrons to enter into hydrogen bonding as in water.

LATENT HEAT OF FUSION

The heat taken up in the change of water from a solid state to a liquid phase with no change in temperature is termed the latent heat of fusion. Specifically, 79.7 cal of heat are required to melt 1 g of ice at 0°C. This is about 15 per cent of the heat necessary to separate the hydrogen bonds in the vaporization process. As heat is applied to ice, the molecules are set into motion.

Increased motion parts the hydrogen bonds until finally the molecular lattice collapses and brings about melting. There is evidence that some few bonds may be broken before melting begins, but even so, only about 15 per cent of the remainder is separated in the actual melting of ice; the fission of the remaining bonds accounts for the high specific heat and heat of vaporization of water.

DENSITY QUALITIES

Most liquids, including water, contract and become heavier with cooling: water is less dense, or lighter, at high temperatures, and becomes more dense as temperature is decreased. Fresh water is unique, however, in that it reaches its maximum density not, as with other liquids, just at the freezing point, but rather at 4°C. As it cools below this point the density decreases, that is, the water again becomes lighter as the crystalline structure develops. Upon freezing, water adds approximately 1/11 to its liquid mass. Thus water pipes burst and ice floats.

Water is one of very few substances whose solid state is lighter than the liquid. This quality is important limnologically, for it accounts for the fact that lakes freeze from the surface downward and that only small bodies of water, and lakes in the coldest regions, freeze solidly in winter; thus in most lakes and streams life goes on underneath the ice—albeit at a reduced rate.

Density is not only temperature-related, it is also a function of pressure, and of the concentration of substances dissolved or suspended in the water. High concentrations of dissolved salts account for increased density of ocean waters and the waters of unique types of lakes such as Great Salt Lake. Organic material such as finely divided detritus and organic silt, being heavier than water, will, in sufficient quantities, increase the density. Similarly, clays and fine muds increase the density of waters in lakes and streams. Viscosity is a measure of the internal, molecular friction of a liquid and is concerned with mobility and flow. In order to flow, the bonding forces holding molecules of a liquid in form must be overcome, and there must be a void in the lattice pattern into which a loosened molecule can move. Compared to many liquids, gasoline for example, water is highly resistant to flow. This resistance is due to the relatively great energy contained in the hydrogen bonds of water molecules.

This quality of water limits stream discharge, and also relates to living habits, morphology, and energy expenditure of aquatic animals inhabiting a dense environment which offers considerable hindrance to movement. From what has been shown of the activity of molecules in relation to temperature, we should expect viscosity to diminish with increase in temperature. Actually, viscosity changes about 3.5 per cent with each degree of temperature, although the relationship is not uniform.

Mass, or weight, of water is also related to density. By general standards, water has considerable weight. Although it varies with temperature, the weight of 1 cu ft of water is generally given as 62.4 lb (998.4 kg/m^3). If determinations are made below the surface, atmospheric pressure must also be considered. At a depth of 100 ft (30.5 m) the pressure is approximatey 58 lb per sq in. (4 kg/cm^2) or about 4 atms (pressure of water increases approximately 1 atm per 10 m depth).

ADHESION, COHESION, AND SURFACE TENSION

Adhesion is the tendency of a liquid to cling to the surfaces of some materials by means of bonds established between the hydrogen atoms of the water molecule and oxygen atoms of the other substance. Cellulose in wood, for example, contains a great number of oxygen atoms and is, therefore, very wettable.

Cohesion is the property of liquids which offers resistance to being pulled apart or to the formation of new surfaces. It has been calculated that a force of 210,000 lb would be required to break a column of pure, perfectly formed water of 1 sq in. (6.45 cm^2) cross section. At the surface of a water mass the force mainly responsible for such resistance is termed *surface tension* and results from the unsymmetrical activity of water molecules at and below the surface.

Whereas the internal molecules are bonded on both sides, those at the surface are attached only to the molecules below, there being none above. A force is thus imparted inwardly giving rise to a surface phenomenon analogous to a taut sheet over the water. The surface tension at the zone of contact between air and water is 73.5 dynes at 15°C. Surface tension varies inversely as the water temperature, and is lowered by the presence of organic substances. The addition of inorganic salts tends to increase surface tension.

SOLVENT ACTION

The phenomenon whereby substances may be dissolved and retained in solution is exhibited to a greater extent by water than by any other liquid in nature. Of the known chemical elements, about 50 per cent have been reported in water and it seems probable that at least traces of all elements may occur. This property is, of course, of prime importance to the maintenance of life in the aquatic environment. Basically two types of processes involving hydrogen bonds are concerned in the dissolving action of water.

One type of solvent process may be described as *inert*, for, by means of hydrogen bonding, the dissolved substances are relatively unaffected by the solvent. Many compounds of ammonia, nitrate, and phosphate, as well as sugars, alcohols, and various organic acids, are bonded to the hydrogen atom of the water molecule through oxygen atoms, hydroxyl groups, or through nitrogen bound to hydrogen.

It should be pointed out that these compounds are some of the important ones involved in energy transfer and storage in biological systems. Therefore, through hydrogen bonding the substances are delivered unmodified and readily accessible, not only from environment to plant or animal, but within the physiological systems of the organisms as well.

The second type of solvent process involves the separation of the electric charge between the hydrogen and oxygen atoms in the H-O-H molecule. Water is characterized by a very high charge separation; thus various salts such as sodium chloride and the salts of potassium are retained in solution.

In still another activity, water has a remarkable ability to ionize dissolved compounds through the separation of the bonds in the H-O-H molecule. This process is due to the high dielectric constant of water. The molecule shows polarity, and total separation of one H from OH results in two charged ions, H+ and OH-. The H+ becomes the hydrogen ion of acids and the OH-becomes the hydroxyl radical of bases. A simple, reversible ionization reaction is that involving the solution of carbon dioxide in water: $CO_2 + H_2O \rightarrow H_2CO_3$, or more completely:

$$O{-}C{-}O + HOH \rightarrow O{-}C\begin{matrix} OH \\ OH \end{matrix} + HOH \rightarrow O{-}C\begin{matrix} OH \\ O{-} \end{matrix} + H_2O^+$$

TRANSPARENCY

Pure water is quite transparent. Since the extent of penetration and absorption of light-wave components, so important in plant and animal relationships, is a function of transparency, this characteristic of water is of great interest to limnology.

Pure water absorbs solar radiation selectively, the major absorption being at the red end of the spectrum. Minimum absorption is in the blue range at about 4700 Angstrom units (A). Approximately 90 per cent of the radiation above 7500 A is absorbed by I m of water.

LIQUID NATURE

The attribute of being liquid constitutes one of the more unique properties of water, for liquids in nature are not common. In addition to water, only one other inorganic substance occurs in liquid form at the earth's crust; this is mercury. It has been suggested that liquid carbon dioxide may be found in quartz, but this is apparently not fully confirmed. Petroleums are liquids, but are of organic origin.

Chapter 13

Solar Radiation and Natural Waters

At this point it might be well to refocus our attention on an idea suggested in the Introduction to our study of inland waters and estuaries. Recall that there appear to be three major themes running through our approach to the ecosystem:

- Fitness of the environment,
- Evolutionary adaptations of the livin components of the systems,
- Traffic in energy.

Of the three, two are more or less directly dependent upon solar radiation in order to maintain their importance and to fulfill their roles in the ecosystem; we refer specifically to the nature of the environment and to energy transfer. As we shall presently see, temperature, as a characteristic of natural waters, is of paramount importance in regulating many chemical, physical, and biological reactions; it is, of course, derived mainly from solar radiation.

Green plants constitute the major source of energy-containing substances available to animals in natural waters. From this relationship stem the intricate food webs involving energy traffic; the ultimate source of this energy is solar radiation.

During the course of a year the amount of radiant energy received by the earth from the sun approaches the staggering figure of 1.3 x 1021 kcal. Some of the solar radiation is reflected from the earth's surface by living features such as vegetation, and by nonliving components represented by land forms, lakes, and seas. A portion of the spectral array is utilized by green plants in synthesizing energy-containing substances necessary for the sustenance of plant and animal life. In addition, heat enters into the development of meteorological processes such as wind and rain which, in themselves, create new habitats and modify old ones.

The known range of radiations in the electromagnetic spectrum extends from extremely short-wave lengths of gamma rays of the order of 0.001 Å (1 A = 1 x 10-8 em) to exceedingly long Hertzian rays of 3 x 1014 Å. Much of this total radiation is distributed outside the earth's surface. The direct solar radiations that impinge on the surface of the earth range from about 135,000 to 2860 A; the most intense radiation, however, extends between 3000 and 13,000 Å.

The peak of radiation distribution is in the bluegreen range, about 5500 Å. Of those radiations that arrive at the water surfaces on the earth, the infrared components (about 0.1 mm to 7700 A) contribute largely to heating. Within a narrow segment of the spectrum lie rays of the so-called light, or visible, spectrum.

These range from 7700 to 4000 A. Below the visible spectrum are found the components of the ultraviolet segment of the solar spectrum, the range being from 4000 to about 2860 A. That segment of the electromagnetic spectrum distributed at the earth's surface.

The earth also radiates, the emissions falling in the infrared range of long waves because the earth is a cool body. The earth's radiations are not entirely lost to outer space because of absorption of some of the radiation by moisture in the earth's atmosphere. At the same time, the atmosphere differentially absorbs solar radiation (on the order of 15 per cent). The net result is the retention of a great portion of the earth's radiation and the transmission of about 85 per cent of solar radiation by atmospheric water Vapour.

The phenomenon whereby the earth's atmospheric water Vapour transmits most of the radiant energy of sunlight while absorbing the radiant energy of the earth, has been called the "greenhouse effect." The principle of the botanical greenhouse is based on the fact that glass of the sides and roof permits the passage of solar energy, but inhibits the transmission of a great part of the "black body" energy given off from the plants and other structures inside the greenhouse.

In other words, the contents of the greenhouse absorb solar energy which passes through the glass; the contents in turn radiate an energy of longer wave length which is not transmitted through the glass. This long-wave heat is therefore retained inside the greenhouse. Actually, greenhouse thermodynamics is more complicated than here indicated, but this simplified statement should point up the analogous roles of greenhouse glass and the layer of atmospheric water Vapour in retaining heat.

Of interest to us in terms of light available for photosynthesis and for heat in maintaining the environment is the mean value of solar radiation falling upon the earth's surface. It has been calculated that an average of approximately 1.5 g-cal/cm2/min of radiation is incident upon the earth at sea level. This value is to be compared with the so-called solar constant of 1.92 g-cal/cm2/min reckoned as the radiation received on the outer surfaces of the earth's atmosphere.

Actually, the light falling upon any area of the surface of the earth may be composed of direct sunlight and scattered light from the sky, or indirect solar radiation. Both of these are subject to considerable variation.

The magnitude of direct solar radiation striking a given point depends upon season, geographical location with respect to latitude, time of day, the angular height of the sun and elevation of the point under observation, and the transmission quality of the atmosphere. The quantity of indirect solar radiation received at a given point is quite variable, but generally constitutes about 20 per cent of the total radiation.

All the light received at the surface of a body of water does not enter the water. A portion of the radiation is reflected, that amount being a function of the angle at which the light strikes the surface of the water and the condition of the surface. From an undisturbed surface, the greater the angle of incidence from the perpendicular, the greater the reflection. When the light rays strike the surface at a very low angle, as much as 35 per cent of the light may be reflected. At high angles the reflectivity is on the order of 5 to 10 per cent. We are concerned limnologically with the light that penetrates the water surface.

One of the conspicuous features of pure water is its high transparency. Pure water does not exist in nature, however; yet the penetration of light into water, and the fate of the spectral components within water, are important in ecosystem relationships, and therefore of primary importance to students of aquatic communities.

The deviation of transparency of natural waters from pure, laboratory water ranges widely, both quantitatively and qualitatively, being determined primarily by dissolved substances, suspended materials, organisms, latitude, season, and the angle and intensity of the entering light. In order to more fully appreciate optical phenomena in, and optical qualities of, natural water, let us first consider some principles derived from laboratory studies of pure water.

The total light entering and passing through a given column of pure water undergoes a reduction in total intensity and change in spectral composition with depth. These processes of reduction and change result from scattering and differential absorption of light by water. The reduction in intensity, or quenching, of the light entering the surface is expressed in the following relationships based upon Lambert's Law:

$$I_z = I_O e^{-nz} \quad \frac{I_z}{I_O} = e^{-nz}$$

In which Io is the original intensity, of the entering light, Iz the measurement of intensity incident at depth z, and e is the basis of natural logarithms; n becomes a constant for a particular wave length, known as the *extinction coefficient*, or the percentage of the original light held back at depth z. The percentage of light passing through z is termed the *transmission coefficient*.

These equations are based on the use of pure water and monochromatic light. Since neither of the conditions occurs in nature, factors which further reduce transmission through the distance Io-1z must be applied for natural waters. Such factors would include, in addition to water, suspended materials and dissolved substances.

It has been demonstrated that in pure water at a depth of 70 m the original intensity of blue light is reduced some 70 per cent, and the yellow component shows only 6 per cent transmission beyond 70m.

The red component is not transmitted beyond 4 m, and orange is quenched in about 17 m. In general, approximately 53 per cent of the total incident light is transformed into heat and undergoes extinction in the first meter of water. The longer wave lengths (red and orange) and the shorter rays (ultraviolet and violet) are reduced more quickly than the middle-range wave lengths of blue, green, and yellow.

LIGHT IN LAKES

In natural waters, the blue segment is transmitted farther than all others. At depths beyond about 100 m blue light is the sole illumination. In highly colored or stained waters, orange and red are transmitted more deeply than other components, but all are rather quickly reduced. In many moderately transparent waters, the greatest transmission is in the yellow.

The most accurate and precise measurements of solar radiation in natural bodies of water are best made with photoelectric apparatus, employing various Colour filters. A photocell contained in a waterproof compartment is connected with a galvanometer and the cell lowered to desired depths. Direct readings of under-water radiation falling upon the cell are made from the galvanometer at the surface.

An American, Professor E. A. Birge, pioneered in the use of electrical equipment for measuring light in lakes. This instrument, developed by Birge and physicists at the University of Wisconsin, was first used in 1912.

Called by Birge a *pyrlimnometer,* it measured the total radiation which impinged upon the surface of a thermopile, the thermal current produced being read from a galvanometer. In some ways, particularly with respect to uniform sensitivity and the expression of energy units, the thermopile remains unsurpassed. But it is costly, and thus the relatively inexpensive photocell has come to be used widely in limnological work. Welch explains in great detail this and other methods of light measurement.

About 1905, Birge was joined in Wisconsin by Chancey Juday and together their researches and reports gave impetus to limnology. The effects of several of the previously discussed factors which relate to transparency of natural waters and comparison with pure water are seen in studying this figure. Total solar radiation is shown as incoming sunlight by the circle with a shaded area at the bottom right. The invisible part of the spectrum, mostly infrared, is represented by shading and includes about half of the entire area.

In this, and the other circles, the portion representing the visible spectrum is divided into sectors proportional to the quantity of each of the six colors measured by Birge and Juday. Beginning at the heavy, horizontal line in the "three o'clock" position, observe that the segments are labeled red, orange, yellow-green, blue, and violet in order of decreasing wave length. The enclosure of the red, orange, and yellow segments in a heavy line, and the arrow at "nine o'clock" will be referred to later.

The circles representing pure water and the four lake examples at given depths are drawn to scale proportionate to the total light present at the depths indicated and to the circle representing incoming sunlight. The unabsorbed light remaining at each depth is given as a percentage outside the circle. Variations in each of the six components of the spectrum are indicated by, the area of the circle segments.

Immediately, apparent is the rate at which transmission decreases with depth, and the variations among the lakes. Similarly, great differences in Colour changes are seen. It is interesting to note that the over-all rate of light extinction and the proportion of the Colour segments for Crystal Lake are not too different from pure water.

In both Crystal Lake and pure water, the yellow-orange-red segments decrease, and the green and blue increase with depth. Accompanying this is a swing of the arrow in a counterclockwise direction.

The data indicate that red waves are absorbed rapidly in the first 3 or 4 m and that with increasing depth the light becomes greenish-blue. In the series for pure water the proportions represent the absorption of direct sunlight; the lake determinations include mixed light from sun and sky. Since mixed light contains proportionately more blue, this light would be transmitted more effectively than sunlight alone in pure water.

As the diagrams for pure water and Crystal Lake would indicate, these waters have no stain. The figures for the other lakes, Silver, Midge, and Mary, indicate increasing amounts of stain and Colour and a proportionate decrease in transparency and light transmission. There is also an increase in absorption and reduction of the green, blue, and violet components. In contrast to pure water and Crystal Lake, the arrow swings clockwise in these stained lakes, especially in Midge and Mary.

The last two are bog lakes and colored dark brown by dissolved and suspended organic materials from the surrounding bogs. In these lakes, orange and brown are the conspicuous spectral components, but nearly all of the incoming light is absorbed in the upper 2 m.

SECCHI DISK TRANSPARENCY

A much older method of measuring transparency of water, and one that still holds considerable value, is the use of the Secchi disk. The method was devised by A. Secchi, an Italian, in 1865. In practice, a white disk, 20 cm in diameter, is lowered into the water by means of a line with measured intervals. The arithmetical mean of the distance at which the disk disappears from view in descent and that at which it reappears in ascent is given as the Secchi disk transparency. In employing this device, care should be exerted to standardize techniques and conditions.

Reflected light from the bottom will introduce an error since the technique involves comparison of the brightness of the disk with the bottom brightness. It has been calculated that the disk disappears at approximately the region of transmission of 5 per cent sunlight.

In very turbid lakes the Secchi disk transparency is quite low. In Lake Texoma, on the Texas-Oklahoma border, values on the order of a few centimeters occur in places. Similarly, artificial farm ponds of the southeastern United States often give readings of less than a meter, particularly, following fertilization when blooms of plankton appear.

Secchi transparency values near 18 m are characteristic of certain lakes in the Convict Creek Basin of California. Generally, values above 30 m are unusual. Crater Lake, Oregon, however, has a transparency near 40 m.

COLOUR AND TURBIDITY OF LAKES

Colour and turbidity are here treated together because of their somewhat similar roles in giving various hues and other optical qualities to lakes; one distinction will be made, however. Both of these qualities determine light transmission in natural waters and consequently "regulate" biological processes within the bodies of water. To varying degrees, both give some qualitative indication of the productivity of the waters when simply viewed from above. The general chemical nature of lakes may often be deduced from the Colour and turbidity of the water. Before proceeding further, we should define our terms in order to avoid confusion.

COLOUR

The Colour that we perceive is made up of unabsorbed light rays remaining from the original entering light, but now passing out of the lake. Completely pure water should absorb all light components and appear nearly black. This is not seen in natural bodies of water, however, for lakes containing little suspended materials usually appear blue.

The blue hue is probably the result of scattering of light by water molecules in motion, the effect being proportional to the fourth power of the wave length of the light components. Since the scattering of light of long wave lengths is less than that of short, the blue predominates.Such hues as we normally see, however, may range from green-blue through blue-green, green, yellow, yellow-brown, to brown.

The vivid, poetic, and not necessarily imaginative impressions such as the "skyblue waters" of Cadman's song, Thoreau's "pure sea-green Walden water," and the "silver" of Seneca Lake as seen by James Gates Percival, all suggest the rich variety of Colour interpretations possible, depending upon one's point of view.

Limnologically, *true Colour* (sometimes called *specific Colour*) of natural waters is derived from substances in solution or from materials in colloidal state.

Apparent Colour, on the other hand, is usually the result of interplay of light on suspended particulate materials together with such factors as bottom or sky reflection. In order to realize true Colour, a sample of the water should be filtered or centrifuged, thereby freeing the water of sources of apparent Colour. The numbers in parentheses near the names of the lakes refer to the Colour of the water in units of the *platinum-cobalt scale* of the United States Geological Survey.

The basic technique involves comparison of lake waters with a series of dilutions of a solution of potassium chloroplatinate (K_2PtCl_4) and crystalline cobaltous chloride ($CoC_{12} \cdot H_2O$). The units are called platinum-cobalt units, based upon 1 mg Pt/l as a standard, and range from zero in clear waters to over 300 units in very dark waters of bog communities.

The Geological Survey set of glass disks corresponding to the platinum-cobalt series is now rather widely used in place of the liquid dilutions. The major classes of dissolved and particulate matter found in typical lake waters may be stated as follows:

DISSOLVED SUBSTANCES

- Proteins and related compounds
- Fats and related compounds
- Carbohydrates and related compounds
- Break-down derivatives of all three of the above

PARTICULATE ORGANIC MATTER

- Living organisms (*Plankton*): mostly microscopic forms.
- Phytoplankton: plant types
- Zooplankton: animal types
- Nonliving particles (*Tripton*): dead organisms, detritus, colloidal substances

The average year-round percentage composition (by weight) of these categories in a "typical" lake is shown in Figure. For any particular lake the bar lengths would doubtless vary greatly; swamp or bog lakes would show a higher percentage of dissolved organic matter; alkaline lakes would be richer in dissolved salts. The important point here, however, is that all elements of these classes, individually or in concert, contribute to the Colour and turbidity of natural waters.

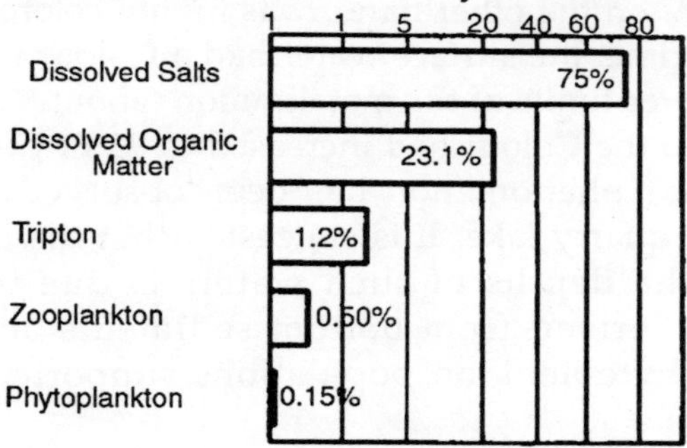

Fig. Year-round Average Percentage Composition of Sestonic Components and Dissolved Organic and Inorganic Material in a Lake.

For the most part, lake colors are determined by the predominating components of the *seston*, the mass of various living and nonliving substances in the water. Plankton algae are most frequently responsible for certain colors; an abundance of blue-green algae imparts a dark greenish hue; diatoms give a yellowish or yellow-brown Colour. Zooplankton, particularly certain of the microcrustaceans, may tint the water red.

Humus often causes water to be green, or yellow-brown, the darkest brown coming from extractives of peat.

Suspended colloidal inorganic substances may account for certain tints. Calcium carbonate in lakes of limestone regions results in a greenish Colour. Volcanic lakes may be yellow-green from sulfur, or red from ferric hydroxide. Generally, a rich, highly productive lake may appear yellow or grayblue or brown due to quantities of organic matter.

Less productive lakes tend toward blue or green. We have just seen that the Colour of lake waters differs widely, depending upon the nature and quantity of dissolved and suspended materials, the quality of the light, and other factors.

It is worth noting also that Colour within a given lake may not be uniform from surface to bottom. Studies in Weber Lake, Wisconsin, revealed that in August the epilimnion was transparent and contained but little organic material; the hypolimnion, on the other hand, was highly colored.

At that time, the surface water had a Colour rating of 12 units, the lower limits of the metalimnion (about 8 m) 81, and at the bottom the Colour had increased to 93 Pt-Co units.

A similar phenomenon has been observed in a New Jersey stone quarry lake. It is suggested that the increase in Colour in the depths of such waters is due to organic substances derived from bottom sediments and also to increased phytoplankton populations supported by such substances.

In a similar vein, the Colour of lakes may change periodically. Seasonal increases in surface runoff contribute great quantities of inorganic and organic substances which, as we have seen, impart various colors. Summer or early autumn production of phytoplankton "blooms" causes a lake to become a "soupy green," which disappears later in the season. Exposure to light causes the bleaching of certain colors in natural waters.

This process also results in variations in the Colour of a given lake, such variations appearing seasonally with cyclic fluctuations in light intensity and angle of incidence.

The bleaching effect should be felt vertically within the lake, depending upon transparency.

TURBIDITY

We have seen that Colour in natural waters is attributable in great part to suspended materials of varying particle size and composition. *Turbidity,* on the other hand, is the term used to describe the degree of opaqueness produced in water by suspended particulate matter.

While the nature of the materials contributing to the turbidity is mainly responsible for the Colour quality, the concentration of the substances, if sufficiently high, determines the transparency of the water by limiting the light transmission within it.

The kinds of materials creating turbid conditions in a given body of water are as varied as the biotic and abiotic composition of the surrounding terrain, inflowing streams, and the lake itself. Substances such as various grades of humus, silt, organic detritus, colloidal matter, and plants and animals produced outside and brought into the lake, are termed *allochthonous*. Turbidity-creating matter produced within the lake is said to be *autochthonous*. Both contribute to the total quantity and quality of lake turbidity.

Although several methods for measuring turbidity have been devised, two are perhaps most widely used today. One technique involves the U. S. Geological Survey turbidity rod, which consists of a marked staff calibrated against known concentrations of a standard material (usually fuller's earth) with a length of platinum wire at one end and an eyepiece at the other.

The end bearing the platinum wire is lowered into the water until the wire vanishes from view. The calibration mark at the level of the water surface gives the turbidity in parts per million (ppm).

Devices such as the Jackson Turbidimeter and the Hellige Turbidimeter make use of light penetration through a sample of water. In the former apparatus, the disappearance of the flame of a standard candle when viewed vertically through a column of water is a measure (ppm) of turbidity. The Hellige instrument compares a vertical beam of light through a column with the Tyndall effect produced by lateral lighting from the same source, usually an electric bulb. An adjustable aperture regulates the light reaching the eye of the observer and the aperture size is calibrated for turbidity and determined in parts per million. For details of these instruments refer to Welch.

Turbidity is not a uniform parameter even within a specific lake. Seasonal increases in stream discharge, for example, may introduce considerable amounts of silt and other sediments and materials, thereby altering the lake Colour and turbidity. With decrease in stream discharge, much of the allochthonous matter begins to settle in the lake basin. The rate of settling is not the same for all classes of materials, grading from sand to colloidal particles.

In natural waters the settling process is complicated by certain attributes of water itself and by the dynamics of water movement, some of which have already been considered. In accordance with Stokes' equation, the rate of settling (velocity of fall of a spherical body through a liquid) is dependent upon gravity acceleration, the radius of the body, the viscosity of the liquid, and the specific gravity of the body and the liquid. Recall that viscosity is related to temperature and that temperature is not necessarily uniform throughout a lake, this being especially important during summer stratification. Turbulence and other water movements contribute to the rate of settling in lakes, particularly during spring and fall overturns.

Some classes of matter which exist in lakes do not settle under normal conditions. These include true colloidal systems or very fine particles of favorable specific gravity, and animal and plant forms capable of modifying their specific gravity and employing other mechanisms (flotation devices, locomotion) to maintain position. The particles in colloidal systems of silt are kept in suspension by the combined effects of turbulence and negative charge on the colloidal particles. It has been observed that the addition of acids to turbid ponds causes flocculation and precipitation of silt particles.

These processes are brought about by neutralization of the negatively charged particles by the positively charged hydrogen ions of the acids. The mechanisms and dynamics of settling and the relationships of particle size to distribution by water movements in lakes are not well known—this in spite of the obvious importance of turbidity in such processes as energy circulation of nutrients, and the effects of sedimentation in lakes.

One most important role of turbidity in the interrelationships of the aquatic ecosystem is the effect, previously mentioned, of suspended materials on transmission of light. This aspect has received considerable attention, especially as it relates to productivity and energy flow within the community. At this point we wish to consider only the effects of turbidity upon light; turbidity as a factor in productivity will be considered later. The damping effect of suspended particles on the transmission of solar radiation, also termed "light-quenching," can be determined effectively from the use of the formula expressing Lambert-Beer's Law,

$$\frac{l_z}{l_o} = e^{-kdc}$$

in which the concentration (mg/l) of light-quenching material c, and the thickness of the column in which light is quenched d, are related to the ratio of observed light I_z to incident light I0 to determine the partial extinction coefficient *k* of suspended particles in natural waters. (This *k* is not to be confused with that representing over-all coefficient in some formulas.)

The parameter I_z may be measured with an underwater photometer. The value of c is the dry weight of suspended matter per unit volume of water after centrifuging. The advantage in the use of LambertBeer's expression over Lambert's Law lies in the recognition of variable *c*; in Lambert's law this factor is included in the extinction coefficient *k*.

The Lambert-Beer expression has been used to interpret data on light quenching in western Lake Erie and other lakes. In the Lake Erie investigation, I_0/I_z was set at 100 (observed light equals 1 per cent of surface light), and d represented the depth at the incidence of 1 per cent of surface light. During the study the highest value for c was found to be 35.9 mg/l with a *d* value of 1.2 m; thus k became 0.107. The highest observed value for *d* was 7.6 and for c was 5.4; a *k* value of

0.112 was then derived. A significant datum was obtained in spring when it was found that the average depth associated with 1 per cent of the surface light was 3.5 m; this degree of turbidity represented a load of suspended matter near 10 mg/l. The rapid absorption of light and the quenching effect resulting from turbidity factors are vividly demonstrated in these figures.

It might appear that Colour and turbidity act together to exert relative effects on light penetration. Such may not be the case in all instances, however. Studies in Atwood Lake, a reservoir in Ohio, have indicated that Colour and turbidity are very probably independent variables exhibiting no interaction with respect to transparency. It was further found that Colour may be the major factor affecting light penetration, except of course during periods of introduction of large amounts of silt during heavy rainfall.

COLOUR AND TURBIDITY OF STREAMS

Basically, Colour in stream waters derives from the same physical laws of light transmission, differential absorption by substances in the water, reflection, and back-radiation as operate in other natural waters. However, as a result of several factors streams generally fail to exhibit the great variety of Colour that may be seen in lakes.

The upper reaches of most streams are characterized by clear waters, at least during the nonflood season. In these regions tile streams lack a true plankton which, as we have seen, may be responsible for certain apparent colors in lakes. Relatively "pure" shallow streams appear clear due, in part, to the fact that light is absorbed quite rapidly in the first meter or so.

In many areas, particularly in the southeastern United States, small, shallow streams which drain "flatwoods" and swamps are often colored a light to dark amber (about 160 on the U. S. Geological Survey Scale). This Colour is probably attributable to dissolved plant substances such as tannin; the streams are typically acid. Similarly, certain small creeks in southern Indiana become 'inky black" in the autumrr due to extractives from large accumulations of leaves in the water. Various extrinsic factors may also account for apparent Colour in streams.

Rich growths of diatoms on rocks on the stream bed may lend to the water a brownish hue; algae may impart a greenish tint, and sulfur bacteria a yellow Colour. Certainly pollution, from a variety of sources, can contribute to both true and apparent Colour in streams and should not be disregarded in field investigations.

COLOUR AND TURBIDITY OF ESTUARIES

The causes and chemistry of colors in marine waters are not thoroughly known. It is likely, however, that dissolved substances impart certain hues or enter into Colour shifts. For example, it is known that water-soluble yellow pigments are common in coastal areas and may contribute to the various shades of green in off-shore waters.

The blue of the sea results, as in inland waters, from the scattering of light by water molecules. Suspended detritus and living organisms give colors ranging from brown through red and green. The Red Sea derives its name from a brownish Colour due to great numbers of particular algae.

As we move into coastal estuaries we find that the brilliant colors of the open seas are generally not apparent. In some estuaries of moderate tide and current action, blue and green hues are noted during the nonflooding periods. Most estuaries, however, are characterized by dark colors resulting from typically high turbidity.

On occasion various planktonic forms become so numerous that they give a reddish or greenish tint to the water. Dissolved materials such as tannic acid delivered by the inflowing stream of the estuary or from local decomposition of organic substances cause estuaries to be light brown.

The major component of turbidity in estuaries is, of course, silt. The volume of silt transported into estuaries by streams fluctuates seasonally, with the maximum discharge taking place during the wet season.

Some materials may be brought in from the sea, but these are usually minimal. In addition to allochthonous silt and other materials, much matter which contributes to turbidity originates from erosion within the estuary itself; the magnitude of autochthonous substances appears to be dependent upon the shape of the basin and prevailing currents.

Throughout the year the amount of materials in suspension in an estuary decreases from the upper reaches to the mouth of the basin, that is, transparency increases downstream. This is due to diminution in velocity and carrying capacity of the inflowing stream current, and to the electrolytic effect of sea-water salts. The latter process involves the coagulation of negatively charged particles of colloidal silt by positive ions of certain metals present in sea water.

Even though there is typically a decrease in turbidity seaward along the axis of an estuary, its waters are decidedly more turbid than the sea. Evidence of this is seen in large areas of highly discolored water lying opposite the mouths of estuaries, contrasting vividly with the clearer sea water. Such masses of estuarine waters often extend many miles seaward or, depending upon currents, for great distances along the coast. The major effects of high turbidity in estuaries are:

- The quenching of light penetration, thereby inhibiting photosynthesis and the production of plants,
- The building of deep zones of mud, silt, other sediments, and detritus.

In many estuaries, notably during periods of considerable stream inflow, light is reduced to 1 per cent of surface radiation at less than 3 m. In certain regions of the open sea the yellow-green components are not diminished to 1 per cent until a depth of nearly 100 m is reached. Transparency decreases shoreward such that in coastal waters the 1 per cent illumination level is generally between 15 and 30 m.

Chapter 14

Thermal Relations in Fresh and Estuarine Waters

From the board and basically ecological point of view, the thermal properties of water and the attending relationships are doubtless the most important factors in maintaining the fitness of water as an environment. We have already considered some of the physical properties which contribute to the total state of the hydroclimate.

Even so, it would be worth recalling the fact that the specific heat of water is among the highest of all substances. This very great absorption capacity accounts for many conducive (and nonconducive) features of the aquatic environment. Remember also that water has the unique attribute of reaching its maximum density at 3.98°C rather than at freezing. These two characteristics will be important in considerations of thermal dynamics in fresh and estuarine waters.

At the beginning of the preceding section we learned of the range of solar radiation that impinges upon the earth's surface. We directed our attention mainly toward those components responsible for illumination so important to photosynthetic activity of plants. At the same time we were aware of the presence of light of wave lengths longer than those in the visible range, namely the infrared rays which range from about 7000 A upward.

We know from Lambert's Law that light absorption by water increases exponentially with the light path. In a column of water 1 m in length 91 per cent of light with a wave length of 8200 A is absorbed; only 9 per cent is transmitted through the column. In 2 m some 99 per cent of this light is absorbed. Over 50 per cent of the solar radiation is absorbed within a depth of 2 m. It is this light with which we shall now be concerned, for the heating of natural waters results from this absorbed radiation. In other words, we should consider *temperature* as an intensity factor of heat energy.

TEMPERATURE AND HEAT IN LAKES

One of the most outstanding and biologically significant phenomena of lakes is found in the relationship between water and temperature as expressed by seasonal variations. In many lakes, these variations take the form of pronounced changes in the over-all thermal structure and dynamics. During winter, the temperature of the water in moderately deep to deep lakes is relatively uniform from surface to bottom; or if ice forms, this colder layer floats on the underlying waters. In spring, circulation and mixing of water results, typically, in a uniform temperature from surface to bottom. During summer, the vertical distribution of temperature may come to resemble that shown in Figure. In this case, the lake is essentially stratified. From the surface to about 15 m, temperature changes little with depth. Between 15 and 30 m, the temperature drops rapidly. The region of the lake below approximately 30 m is rather uniformly cool. In the fall, the summer condition is broken up by circulation and mixing similar to that of spring, resulting once more in uniform temperature of the water.

How do these conditions come about? The answers are found in light absorption, heat dynamics, density phenomena, and wind action. These are best appreciated by considering in more detail the *annual temperature* cycle of a lake.

SEASONS IN LAKES

In winter, as shown for March at the extreme left of Figure, ice at near 0°C covers the surface of the lake. Recall that ice floats because it has a density less than that of water. Below the ice the temperature to the bottom is relatively uniform and, as suggested by the temperature of 1.5°C at about 15 m, gives evidence of being rather thoroughly mixed-the mixing having occurred in the fall before the ice formed.

The density is nearly uniform below the ice. Because of the low angle of the winter sun and the shading of the water by the snow and ice cover, photosynthesis is inhibited. This, together with respiration of organisms and lack of oxygen replenishment from the atmosphere, results in low oxygen content of the water. Indeed, by late March oxygen is nearly depleted toward the bottom.

In the spring, a higher sun and increased day length brings about the melting of the ice. As the surface waters warm up to 4°C and become more dense, a slight, and temporary, stratification exists which sets up convection currents. These currents, aided by wind, serve to mix lake water throughout until it is uniformly at 4°C and at its maximum density.

Note from the vertical isotherms in Figure, that in Lake Mendota the mixing continued into early May even while the water was warming to 10°C. It is apparent that the amount of dissolved oxygen had increased throughout the lake. This vernal process of mixing has been termed the *spring overturn, or spring circulation period*.

Now we have seen that heat is absorbed very rapidly in the surface waters and, therefore, cannot be responsible for heating the deeper waters. Thus some other agent must contribute to the circulation of warm waters into the deeper regions. *Heat transfer in lakes is accomplished mainly by winds*. It is only by this force that mixing can continue beyond 4°C. Winds blowing across the surface of water set up a current resulting from frictional differences between moving air and water. Upon reaching the shore this current moves downward and across the bottom, thereby setting up the spring circulation.

As summer approaches, the weather warms, the longer days mean longer periods of insulation, and the brisk spring winds subside. Under these conditions the surface waters warm rapidly, expand, and become lighter than the lower waters. Although the wind may continue to blow, its contribution to mixing of lake waters diminishes, for now the thermal density gradient opposes the energy of the wind. With the progression of the summer season the resistance to mixing between two layers of different density (resulting from increased temperatures) becomes greater than the force of winds—the significant density differences having been built up during periods of summer calm.

Reference to Figure reveals that Lake Mendota became thermally stratified during June and July. In late July, the uppermost region of warm, homothermal water extended downward some 8 m. This region is termed the *epilimnion*. Below the epilimnion there developed a zone of rapid drop in temperature with depth.

This zone, only 2 to 3 m thick, is the *thermocline*. In modern limnological parlance the thermocline is defined as "the plane of maximum rate of decrease in temperature." The *zone* of rapid drop in temperature, including a gradient on either side of the thermocline, has recently been termed the metalimnion.

The condition of thermal stratification is very stable. The metalimnion constitutes an effective barrier between the epilimnion and the hypolimnion.

Currents stimulated by wind, as well as convection currents derived from cooling at the surface, are limited in depth by the metalimnion. This means that heat and nutrients present in the region of maximum light are prevented from mixing throughout the body of the lake. Similarly, hypolimnetic water substances and many nonmotile organisms and those of limited mobility are restricted to the region.

For photosynthetic organisms this state of affairs is particularly critical. Depending upon the depth of the metalimnion, of course, photosynthesis is essentially inhibited in the hypolimnion.

This means that oxygen available for animal respiration is reduced, carbon dioxide is increased, and that microscopic algae (phytoplankton) as food for herbivorous animals is scarce. Indeed, it has been found in many lakes, Fayetteville Green Lake, New York, for example, that minute animals (zooplankton) eat great quantities of purple bacteria which are abundant in the deep, oxygenless waters. The "impedance" to over-all lake current flow presented by the metalimnion has been called the *thermal resistance to mixing*. What, we might ask, is the character of this resistance, and how is such a barrier maintained? The answer to the first question is found in the density-temperature relationships mentioned briefly a few paragraphs back.

In Figure we see that between depths of about 15 m and 30 m the temperature declines from near 20°C to approximately 8°C, or about 12°C in 15 m. Recall that the density of water is dependent upon temperature, and that the difference in density per degree change is fairly great.

For example, as water warms from 4° to 5°C the density changes 8 x 10^{-5} gm mass/ml. Therefore within the rapid and wide temperature change in the metalimnion a considerable range of density also exists. Consider also the density difference between the temperature at the epilimnion-metalimnion interface and that at the hypolimnion-metalimnion interface in Figure.

This difference represents a strong value of 0.00164 gm mass/ml. Where temperature-depth data are available a graph of the *relative thermal resistance* can be drawn. This parameter represents the ratio of the density difference between water at the upper and lower faces of each stratum to the difference between water at 5° and 4°C.

We might conclude then, that resistance to mixing presented by the metalimnian region is primarily the result of temperature decrease with depth and the related density differences within the zone. Now, what perpetuates the metalimnion and its resistance to mixing? We have already learned that the epilimnion is a region of relatively free circulation and considerable turbulence, and further, that the metalimnion acts to limit the downward extent of water motion. As wind-blown surface currents move against the shore of the lake they are deflected downward and encounter the metalimnion.

The metalimnion barrier in turn causes the current mass to spread out horizontally at the upper zone of the metalimnion. Each increment of warming of surface waters in the summer adds a proportionate measure of relative thermal resistance between the epilimnetic and metalimnetic waters. Thus the barrier effect is somewhat self-perpetuating in the sense that density differences between the two regions are increased as the continually warning waters circulate toward the cooler metalumnion.

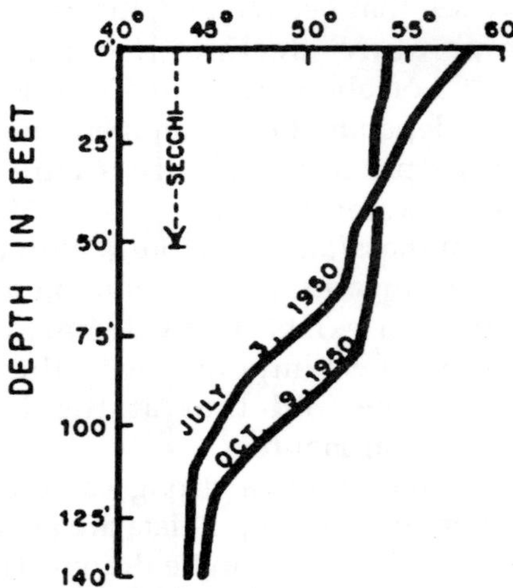

Throughout the great range of lake morphology, climate, and topography, the various attributes of the thermoclune are subject to considerable variation. The thickness of the metalumnion may fluetuate with season, becoming thinner as summer progresses. This is shown rather clearly in Figure, but in other lakes the effect may be more pronounced, the thickness ranging from about 12 m in early summer to around 5 m in late summer.

Recently a very thin metalimnion of some 2 to 5 mm in barber's Pond, Rhode Island, was reported. Across this layer, visible to underwater observers, the temperature decreased almost 8°C.

The depth of a metalimnion is determined by a number of influences such as length and time of seasons, and by basin morphology. Within a particular lake the metalimnion usually becomes deeper in late summer.

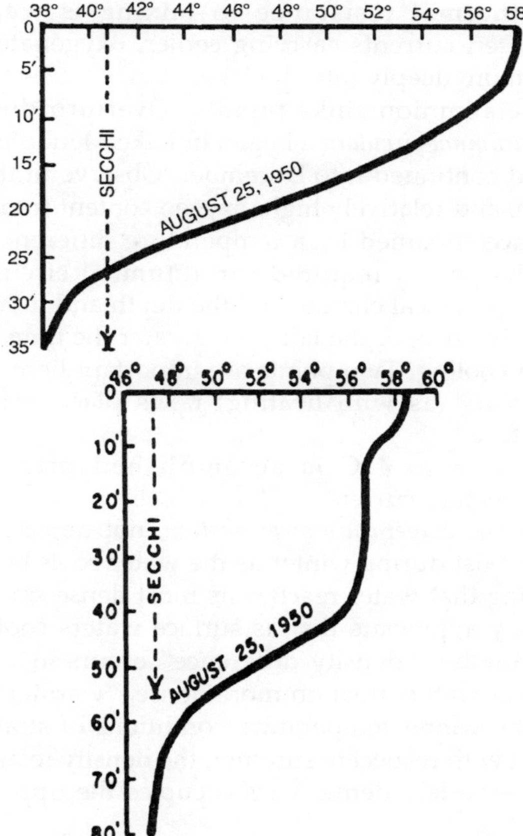

Fig. Temperature Profiles for Three Lakes of Various Depths

Local variations in metalimnion level in response to wind action and to the temperature of incoming waters may be expected. One additional point should be considered here; smooth temperature curves are not always obtained. Typically a late-developing stratification results in the formation of a thick epilimnion and more than one metalimnion.

With the approach of autumn, the angle of incident light decreases, day length shortens, and cooling begins. This is to say that the lake loses heat faster than it is absorbed. As the cooling extends toward the deeper regions of the lake, the density differences between isothermal strata become less.

Thus the thermal resistance to mixing is weakened. Windgenerated currents carrying cooler, oxygenated water reach ever more deeply into the lake.

The metalimnion sinks rapidly. Overturn during the period of *autumnal circulation* began in Lake Mendota in early October and continued into December. Observe in the figure the uniform, and relatively high, oxygen content from surface to bottom accompanied by a temperature difference of less than 10. The period required for autumnal circulation is dependent upon local climate and the depth and morphology of the lake. The deeper the lake, the greater the time required for uniform cooling. Two points are important here:

- Cooling (as with heating) takes place only at the surface,
- Cooling to 4°C is accomplished primarily by convection currents.

An *inverse temperature stratification,* not depicted in our figure, may exist during winter as the water cools below 4°C. Remembering that water reaches its most dense state at 4°C, we can easily appreciate that as surface waters cool beyond that point another "density difference" occurs in which the lighter, cooler waters float on more dense, "warmer" waters. Although the winter temperature conditions of stratification are reversed with respect to summer, the density relationships are similar—the less dense mass occupies the upper part of the lake.

The inverse stratification is most pronounced following the formation of the ice cover. Immediately below the ice (0°C) the temperature of the water rises sharply to near 4°C, or whatever the temperature of the water body. Inverse stratification is usually of rather short duration and, indeed, may not occur every year in a given lake. The period of *winter stagnation now exists.*

The sources and processes involved in the heating of lake waters beneath ice have received considerable attention. It appears, however, that heat from direct solar radiation through the ice, and heat derived from bottom muds are the major sources of energy for warming the lake.

The heatdistributing mechanisms are apparently density currents of two types:

- The temperature-dependent currents which are derived from heating of water through the ice in the shallow zones and which flow into the deeper parts of the lake,
- Chemical density currents derived from dissolution of bottom substances in water; more dense by virtue of the dissolved materials, these currents may also transport heat.

THERMAL CLASSIFICATION OF LAKES

As a result of extensive studies of temperature phenomena, heat, and stratification, since the time of Forel, a great store of terminology and schemes of classification of lakes based on thermal characteristics has accumulated. In an attempt to keep terms and classifications at a minimum we have adopted a scheme proposed by Hutchinson.

By no means universally applicable, by virtue of the variable nature of lakes the system is useful and points to the more modern concepts and approaches to thermal properties and dynamics in lakes. Earlier terms and classifications, certainly not useless, are to be found in older texts and other works on ecology and limnology.

The following types are proposed for lakes occupying basins of sufficient depth to allow for stratification mixing ("mixes"), and the formation of a hypolimnion. Consideration has been given to altitude, geographical location with respect to latitude, and to depth of the basin.

AMICTIC

Lakes insulated and protected by ice cover from outside influences of weather and other factors; poorly known lakes of the Antarctic and high altitudes.

Cold Monomictic

Lakes of the polar regions in which the waters at any depth never exceed a temperature of 4°C; ice-covered and exhibiting an inverse temperature stratification in winter; one mixing ("monomixes") at temperatures not greater than 4°C in summer.

Dimictic

Lakes in which two circulations take place each year in spring and autumn; thermal stratification is inverse in winter and direct in summer; typical of lakes in the temperate zone and in higher altitudes in subtropical regions.

Warm Monomictic

Lakes of the warmer latitudes in which the temperature of the water never falls below 4°C at any depth; one circulation each year in winter, directly stratified during the summer. Lake Providence, Louisiana, stratifies from May through September, and circulates continuously from October to April.

Oligomictic

Warm lakes in which the water temperature is considerably higher than 4°C; circulation occurs rarely at irregular periods; found typically at low elevations in tropical zones.

Polymictic

Lakes in which mixing is continuous but at low temperatures, usually just over 4°C; characteristic of high mountains in the equatorial regions. Stratification does not develop because of heat loss to a relatively uniform environmental temperature.

HOLOMIXIS AND MEROMIXIS

With respect to circulation, or mixing, generally, most lakes are said to be holomictic ("wholly mixing"); that is, if circulation takes place it is complete and extends the entire depth of the lake. In these lakes the temperature of the hypolimnion usually decreases uniformly to the bottom. However, some lakes are known in which the summer temperature curve resembles that of a holomictic lake except that temperature increases slightly through a considerable distance. How is it that warm water is lying underneath cold water? What of the density relationships? This bottom water is not less dense than the upper water.

It has been found that the bottom stratum usually contains large quantities of dissolved salts, and these increase the density sufficiently to prevent mixing with the upper layers. We have, then, a lake in which circulation is not complete; such lakes are called *meromictic* ("partly circulating"). The bottom, noncirculating layer is called the *monimolimnion,* and the density gradient becomes a *chemocline.* The waters above the chemocline in which thermal stratification such as in holomictic lakes can occur is termed the *mixolimnion.*

A lake in western Austria (Längsee) possesses a monimolimnion that has not mixed with the upper waters for over 2000 years. The monimolimnion contains a heavy concentration of salts derived from the sediments by biochemical processes. Heat in the bottom layer is apparently gained from bacterial activity and from insolation. Since there is no circulation by currents, heat is lost mainly through conduction. The annual temperature cycle of a meromictic lake is shown in.

HEAT BUDGETS AND LAKE STABILITY

Having considered the annual temperature cycle in a typical lake of the temperate zone, as well as some unusual cases of stratification, let us now give attention to some of the broader aspects of heat and heating. As stated earlier, temperature is the expression of the intensity phase of heat energy, while heat, as such, is an energy factor.

For purposes of comparison of like qualities and expression of the nature of the lake waters as environment with respect to organisms, the calculation of heat income and heat budgets becomes a useful tool. *Summer heat income* is the quantity of heat delivered to a lake necessary to warm the waters from the homothermal spring condition of 4°C to the summer maximum. Although direct solar radiation contributes essentially all of this heat, the distribution of it is accomplished by winddriven mixing of waters of differing densities. The work of the wind in summer heat income is done against gravity because of density differences arising from heating of surface layers by direct insolation.

The *annual beat budget* takes into account the total quantity of heat taken into the lake to warm the waters from the lowest temperature of winter to the maximum summer temperature. Either parameter can be determined in absolute terms such as calories, or gram-calories. Usually, however we wish to compare heat characteristics of several lakes, in which case it is well to introduce area. The result is then expressed in calories per square centimeter of water surface. The heat content per unit surface area of a lake at any given time can be determined by totaling the product of temperature times volume for declared depth intervals, and then dividing by the surface area of the lake.

The gain in heat content per unit area beyond the hemothermous 4°C state is the summer heat income, and is more a function of mean depth than of temperature differences. The annual heat budget may be derived in the same way, except that the lowest winter temperature is substitute for 4°C. Table gives morphological and thermal data for five lakes in the Convict Basin of California, and shows the relationship between depth and annual heat budget.

Comparisons of annual heat budgets of lakes must consider depth differences. Shallow lakes are not able to store all the heat transmitted into them. During the period of maximum temperature, the lakes become vertically homothermous and heat is transferred and lost to bottom sediments.

Within temperate regions the annual heat budgets of typical lakes range from about 30,000 to 40,000 cal/cm^2. Much of this heat (33 to 75 per cent) enters the lake after the spring circulation period, and work energy is involved in the circulation of the heat to lower lake levels. lake Michigan, with an annual heat budget of 52,400 cal/cm^2, and Lake Baikal, whose annual budget amounts to 65,500 cal/cm^2 are the highest known. The summer heat income for Lake Michigan is 40,800 cal/cm2, and Lake Baikal receives 42,300 cal/cm^2 between spring isothermal 4°C and summer maximum. To bring about the thermal condition shown in Figure for Lake Mendota an annual budget of about 23,500 cal/cm^2 and a summer income of some 18,240 cal/cm^2 are required.

Table: Major Physical Features of Certain Lakes

					Temperature, °C	Heat Intake, gm cal/cm2		
Lake	Elevation	Surface Area (ha)	Mean Depth (m)	Transparency (m)	Bottom to Surface	Lake Mean	Summer Heat Income	Annual Heat Budget
Convict	2275	68.6	26.4	15.3	6.6-14.7	10.9	18,173	22,984
Mildred	2970	4.3	8.1	12.0	10.0-12.5	11.1	5,822	7,298
Witsanapah	3228	1.8	8.4	9.0	5.5-10.0	8.5	3,827	5,358
Edith	3030	7.3	17.7	4.1-14.1	13.8	8.7	8,601	11,826
Dorothy	3102	43.8	40.8	20.1	4.4-14.7	8.8	17,967	25,402

In view of the importance assigned to the work of wind in distributing heat in a lake, we should question how much work a wind of a given velocity can do and how much work is required to mix a given quantity of heat. As Hutchinson has pointed out, little work has been done on the problem, and our knowledge is therefore meager.

However, some data from various sources as compiled by Hutchinson will serve to convey a general idea. With respect to the first question, Langmuir has stated that winds of velocity 300 to 700 cm/sec developed forces of 0.65 to 6.3 dynes/cm^2 on lake surfaces. The summer heat income of Green Lake, Wisconsin, is 27,316 cal/cm^2 and the heating period is 122 days. Calculations show that the mean rate of work directed toward heating the waters is 0.02 dynes/cm^2. These data indicate that the work of the wind in summer heat distribution is very small in comparison to the force of wind on the lake surface. The loss of heat during the period of cooling from the summer maximum temperature to the winter minimum must be of the order of the annual heat budget. This loss from the lake is mainly through radiation to the surroundings. Cooling by evaporation also expends heat.

The waters of effluent streams remove considerable quantities of heat from lakes, especially from impoundments where warm, surface waters are removed while cooler waters remain behind. Heat exchange processes in a lake and impoundment are shown in Figure. From the foregoing considerations we sense that a certain quantity of energy is expended in developing summer stratification in holomictic lakes. We have also made reference to the stability of the stratification and to the resistance offered by the lake to mixing. This, in essence, is one definition proposed for the term *stability*.

More properly, *stability of stratification* is that energy of resistance which the lake offers to oppose upset of density stratification. As warm, less dense water comes to overlay cool, more dense water during stratification the centre of gravity is shifted downward.

Stability then becomes a measure of the amount of work required to raise the centre of gravity or to displace it to its original position. Stability decreases in autumn as the thermocline sinks below the centre of gravity and the waters above and below this centre attain similar densities. The greatest stability is probably reached just prior to maximum heat content in summer.

Table. Summer Pond-water Temperatures

	°C
At surface of Mat	21 to 22
2.5 cm Below Mat Surface	23 to 24
5.0 cm Beneath Mat	25
7 to 8 cm Beneath Mat	21 to 23
15 cm Beneath Mat	20

While we have been considering temperature and heat relationships in relatively large bodies of standing waters, we should not lose sight of the fact that pools and small ponds also exhibit responses to seasonal and daily temperature influences. Such responses often limit activity of some organisms and should certainly be studied in limnological investigations of small situations. "Microthermoclines" and odd temperature stratifications exist in certain small bodies under given conditions. A reversed stratification has been shown to exist at times in small Burt Pond, Michigan.

The data for this pond are given in Table and indicate, among other things, how rapidly the temperature of a small body of water responds to atmospheric conditions. It is suggested that the reversed stratification results from heating of the surface by direct radiation in the morning, followed by cloudiness and winds which cool the surface in the afternoon.

THERMAL PROPERTIES OF STREAMS

The basic thermal characteristics and processes in the water in streams are not different from those of lake waters. Penetration of light, absorption, specific heat, and other factors attributable to the nature of the water molecule operate in determining the fitness of the stream as an environment just as in lakes. As an environment, however, a stream presents a considerably different set of temperature conditions.

These conditions derive from variations in velocity, volume, depth, substrate, cover, water source, and a number of additional features operating seasonally, daily, and even longitudinally along the stream course at a given time. Recall that we have termed a stream community an "open system" by virtue of the considerable land-water interrelationships; some of these influence temperatures.

The major factor in the warming of stream waters is direct solar radiation. A series of temperature curves recorded over several successive days of cloudless skies followed by an overcast day will show rather regular fluctuations from late afternoon maxima to early morning minima during the cloudless period, but the cloudy-day curve will be flattened. Additional evidence for the importance of direct insolation is found in the frequent occasions when the temperature of the water exceeds that of the air.

This usually occurs on very clear days of intense sunlight. The limit of nocturnal cooling is, of course, regulated in small, shallow streams by atmospheric temperatures. In larger, deeper streams it depends upon the rate of heat loss before warming begins again. Additional heat may be gained from bottom sediments in streams, but little data are available to show the extent of such heating.

On the other hand, the temperature of stream water is a measure of the actions and interactions of a wide variety of factors. Consider a stream rising in a rocky, wooded highland and flowing down over the coastal plain to the sea. In the upper reaches the waters are cooled by the substrate, by the shading provided by vegetation, and possibly by the entrance of spring-fed tributaries.

Turbidity may be quite low. As the stream approaches the lowlands it becomes wider and deeper, and more water is exposed to direct sunlight. This, and increased silt content which absorbs considerable heat, result in the development of a segment quite different with respect to temperature from the upper reaches.

Depending upon size and origin of streams, their diurnal and seasonal temperatures follow atmospheric temperatures more closely than do those of lakes. Diurnal temperature variations at a given point in a stream may be related to two major sets of factors:

1. Conditions at that point,
2. Conditions upstream from the point.

Under (1) we should consider velocity and discharge, season and hour, and the daily range of fluctuations of air temperatures at the point. Factors pertaining to (2) include the nature of upstream environment, substrate (and impoundments, if any), atmospheric conditions and temperatures upstream, and distance and time of flow from critical upstream situations.

With respect to size and water temperature fluctuations, we can conclude that the smaller the stream, the greater the temperature variations and the more rapid the response to environmental fluctuations. Furthermore, these fluctuations exist throughout the year, but are minimized beneath ice cover. The range of daily variation of water temperature is maximal when there is the greatest differential between mean diurnal air temperature and mean water temperature.

With increased volume and turbidity the range of fluctuation diminishes. Because of the wide range of temperatures exhibited by small, cold streams, a single temperature reading may not at all approximate the average daily figure.

The source of the stream's waters and the nature of the drainage pattern in some instances determine the thermal properties of the stream. Many spring-fed streams and those originating from surface discharge of subsurface aquifers are essentially thermostatic. Lander Springbrook, New Mexico, is a small rheocrene ("flowing") spring. Seventy meters below the source the annual variation in water temperature is 4°C; air temperatures fluctuate from about 14.3° to 33°C.

The annual variation of water temperature of Silver Springs, Florida, a large spring run (daily discharge about 600 million gallons mainly from aquifer openings) varies no more than 1° throughout the year. Diurnal fluctuation in Silver Springs is also of the order of 1°. These spring-fed streams do not receive significant quantities of surface water which might otherwise bear on their thermestatic qualities. Surface-fed streams typically show wide seasonal fluctuations corresponding with atmospheric conditions.

Because of turbulence and the shallow nature of most streams, thermal stratification is not generally an attribute of streams. When stream waters stratify, the process usually takes place in pools along the stream course. In some instances the stratification is a result of inflow of cool spring water at the lower levels of the pool. In Jack's Defeat Creek, Indiana, a pool was found to be stratified, the temperature decreasing 7.5°C from surface to bottom at about 62.5 cm.

Other qualities, including free carbon dioxide, dissolved oxygen, and alkalinity were also stratified. Inflow of spring water possibly accounts for the local stratification in this case. If warmed upstream waters are gently introduced into a cool pool of water, the inflowing, less dense waters are apt to flow over the more dense pool waters and on downstream without considerable mixing.

One word on the effects of human endeavors, such as

agriculture and industry, on stream temperature and ecology needs to be interjected here. We know that water taken from a stream and used in irrigation is warmed in the process and, upon being dumped back into the stream, increases the downstream temperature for some distance. Similarly, water used in cooling various industrial installations warms the stream.

In addition to the simple warming of the stream which, as such, may exceed the temperature tolerance of some of the organisms, the increased heating also decreases the oxygen retention capacity of the water, thereby affecting certain organisms and stream metabolism. Reservoirs, whether for water supply or hydroelectric power, exert considerable influence on the quality of stream water below the impoundment.

As spring temperatures of streams approach the favorable point for spawning of warmwater fishes, sudden discharge from the lower levels of an upstream impoundment cools the stream and inhibits reproduction of the fishes. Conversely, the influx of warm, surface waters from an impoundment into a cool trout stream has deleterious effects on the fishes and the stream community. Much precise work needs to be done on these aspects of stream relationships. The results of such research could go far in enlightening the agencies concerned with dam-building projects.

HEAT AND TEMPERATURE IN ESTUARINE WATERS

Before considering some of the more general thermal aspects of the estuary as a body of water, we should give attention to the effect of increased concentration of salts upon certain basic properties of water. Recall that pure water has a specific heat of 1. As the concentration of dissolved salts increases, the specific heat decreases. For sea water at 17.5°C and salinity, of 35‰ (parts per thousand) the specific heat is 0.932. This suggests that less heat is required to warm a given volume of salt water than to warm the same amount of fresh water. Whereas for pure water density is a function of temperature alone, being maximum at 4°C, the presence of salts depresses the temperature of maximum density in sea water.

At a salinity of 24.70‰ maximum density occurs at the freezing point,-1.33°C. At salinities higher than 24.70‰ the temperature of maximum density is below the freezing point. Unlike pure water, sea water at high salinities undergoes regular increase in density as it cools to freezing. Generally, the freezing point of sea water varies inversely as the salinity. At a salinity of 10‰, commonly experienced in estuaries, the freezing point is near-0.5°C; at 35‰ sea water freezes at about-2.0°C. The complexities in estuarine thermal dynamics are easily appreciated when we remember that within a few miles the salinity in an estuary may grade from near 0‰ to 25‰ or higher.

The heat content of estuarine waters is derived mainly from solar radiation. These waters are directly heated *in situ* as they occupy the estuary basin. Heat is also received indirectly from inflowing stream water and from tidal flow from the sea. The temperature of the estuary is, therefore, primarily a function of the temperatures of entering streams and the sea together with tidal stages.

If the estuary empties into a relatively deep sea, the seaward temperature is apt to be more stable than in the upper zone where the stream temperature may fluctuate widely. On the other hand, estuaries of spring-fed streams flowing into shallow sea zones may exhibit more stable features in the headward regions.

In view of the shallow nature of most estuaries we should expect considerable diurnal and seasonal fluctuations in surface temperatures. In the East Bay, an arm of the Galveston Bay system of Texas, during July, surface water warms from an early morning temperature of about 27.0°C, to 33.0°C in late afternoon; atmospheric temperatures during the same period range from near 28.0° to 34.0°C. Seasonal temperature fluctuations depend upon latitude and a number of local factors such as water source, basin morphometry, winds, and tides.

In Apalachicola Bay, Florida, the annual temperature

range is of the order of 25°. Temperatures in the Sheepscot estuary of Maine range from about freezing to near 25°C in the upper reaches of the estuary, while in the lower region the range is about 15°. These temperatures are determined primarily by atmospheric and climatological conditions. A number of instances have been noted, however, in which sudden changes in the pattern of oceanic currents result in the influx of cold masses of water into an estuary, thereby drastically reducing water temperature.

Another factor that affects temperatures in estuaries is the flooding of marshes and mud flats during high tides. In warm, humid regions the surface of exposed marshes may, during the summer, become quite warm. As the rising tide covers the marshes with estuarine waters, heat is imparted to the waters. In other regions the exposed muds may become cooled through evaporation and thus lower the temperature of incoming water. Studies in the Elkhorn Slough estuary of California have shown that mud gains heat much less rapidly than the over-lying water, even in a depth of only 15 cm.

Temperature distribution in estuaries is largely a function of depth together with the relative effects of stream inflow and tidal exchange. In a shallow, mixing estuary the waters tend toward vertical homothermy. Longitudinal temperature patterns vary seasonally.

Depending upon the morphology (surface-volume ratio) of the stream basin and the volume and rate of discharge, the upper waters of the estuary may be cooler in winter and warmer in summer than the lower estuary waters. From the mouth of the Patuxent River of Chesapeake Bay to a point about 74 km upstream, the average surface temperature from March to July grades from about 23° to 21°C; bottom-water temperatures similarly decrease on the order of approximately 3°. From September to January, both surface and bottom temperatures increase slightly over 1° from the bay upstream.

Where conditions of climate and sufficient depth are met, estuaries may exhibit a vertical temperature gradient. In summer the surface waters are usually warmer than the underlying layers. With the cooling weather of autumn, the upper waters cool more rapidly than the deeper masses, resulting in overturn and mixing. In midwinter the temperature of the surface waters may hover near that of maximum density, thereby giving rise to convection currents which continue to circulate the waters.

Increasing insolation and winds in spring warm the surface, and a temperature gradient follows. Density stratification derived from cold or warm fresh water flowing over the more dense salt water also contributes to vertical temperature differences in estuaries. Such a stratification may be recognized by salinity measurements from the surface downward.

Chapter 15

Natural Waters in Motion

In the smallest pool and the largest lake or stream, movement of some form and degree takes place. Motion in such waters derives from temperature and density differences, gravity, and wind acting together or singly and to varying extent.

Directly or indirectly every facet in the aquatic community, be it physical, chemical, or biotic, is in some way affected by movement within the body of water.

In this chapter we wish to single out some of the more conspicuous forms of motion in lakes, streams, and estuaries and to consider them as basic features of natural waters and as physical factors determining the fitness of waters as an environment for habitation by living organisms.

MOVEMENT OF WATER IN LAKES

As a general rule, the water of a lake basin is partially or wholly in motion. It is this movement, derived from either internal or external forces, or both, that is responsible for the circulation of heat, dissolved substances, and some organisms in lakes.

Turbulent flow is characteristic of the movements of lake water; this results from the action of molecular systems imparting irregular direction and velocity to water mass.

Turbulent movement is then incorporated into the more conspicuous forms of motion in lakes. The larger movements of water are called *current systems,* and are frequently defined as two types: nonperiodic, or arhythmic; and periodic, or rhythmic.

NONPERIODIC CURRENT SYSTEMS

Nonperiodic systems are those often termed simply, "currents," implying a unidirectional flow of water. This movement may be caused by differential heat distribution within the lake, the passage of stream waters through the lake, and by winds. Broadly speaking, nonperiodic systems are produced and maintained by external forces.

Wind is doubtless the major external force acting to set lake waters in motion. Sustained winds blowing across the surface serve to pile up water at the down-wind end of the basin. The result is lowering of the water mass in another part of the lake. As the winds subside, the piled-up water begins to flow along a gradient, or down the slope, so to speak. Eventually the over-all lake level reaches equilibrium and the current system comes to rest. We have seen that the action of wind is also important in the development of summer stratification of lakes.

A phenomenon familiar to all who have traveled the ocean or large lakes is that of *wind streaks*, or those linear accumulations of foam, floating debris, or oil. Irving Langmuir, better known for his work in electricity and related fields, demonstrated that surface flow resulting from wind drift is not an even and uniform movement of water containing variable velocities, but rather possesses a definite pattern.

Water flow in wind drift is in the form of helices lying parallel to one another oriented in the direction of the wind with the direction of the helices alternating clockwise and counterclockwise, and the effect of streaks is derived from the accumulation of materials in zones of convergencies between clockwise and counterclockwise helices. The width of the zone of convergence varies seasonally, being wider in the late autumn than in spring. It appears that the downward vectors of the helices serve to transfer momentum from the surface into the deeper regions of the lake.

In large lakes the effects from stream inflow, wind, and earth's rotational forces often combine to produce a pattern of large current swirls. The surface currents of Lake Constance, in Europe, have been studied rather intensively, especially with relation to the path of the Rhine River through the lake. As a result of geostrophic forces and the shape of the basin, the Rhine waters flow across the upper end of the lake and follow the northern shore. The river current, together with wind patterns, apparently produce the large swirl between Langenargen and Rorschach.

The influence of the Rhine on the currents in Lake Constance serve to point up another kind of nonperiodic current in lakes; such may be termed *density currents*. These currents result from the passage of river water of a given density through a lake of differing density. As suggested previously, the route of the stream water is determined mainly by geostrophic forces, the shape of the lake basin, and local meteorological agents. The depth at which the river water flows is a function of density differences.

If the density of the river water is greater than any of the lake waters, the stream flows over the bottom of the lake. Similarly, if the lake is generally more dense than the stream, the latter flows at the lake surface. During the summer months the water of the influent is usually cooler than the lake surface, and therefore more dense.

Thus the river water flows downward until it meets a region of greater density, whereupon the river flow becomes horizontal. The "waterfall" at the mouth of the Rhone in Lake Geneva is familiar. In thermally stratified lakes, the river flow is usually above the hypolimnion, thus the lower lake levels receive little of the value of enrichment by the inflowing stream.

In Norris Lake, a Tennessee Valley Authority impoundment, there are indications that density currents move through the lake as the water is drawn down for hydroelectric power. These currents may be poor in dissolved oxygen content due to organic decomposition, causing movement of fishes to avoid the oxygen minima.

Density currents may be important in ponds and small lakes, and probably relate to distribution of materials beneath ice cover in winter. In Lake Mead, formed by Hoover Dam on the Colorado River of southwestern United States, a density current resulting from increased silt load forms during the summer. This *turbidity current* may flow into the lake below the level of the river density current.

One more aspect of nonperiodic movement of water should be considered, that of *turbulence* and *eddy effects*. Although the mechanics of these phenomena are best known from laboratory studies, the applicability of the principles to limnology is vast and deserving of more intensive study in the future.

To anyone who has watched a stream plunging through riffles or along irregular shores, the intense turbulence and the resultant eddy systems are obvious. Within lakes, however, such turbulence is reduced and certainly less conspicuous, albeit present as the result of currents such as those previously described and to be discussed later. The great importance of eddy effects is in heat transfer within the water mass (*eddy conductivity*), in the diffusion of dissolved materials (*eddy diffusivity*), and momentum transfer (*eddy viscosity*).

We distinguished laminar and turbulent flow with respect to streams. It is agreed that laminar flow, wherein sheets of liquid move uniformly without local variations in velocity (but with velocity differences from "sheet" to "sheet"), does not normally occur in nature.

The more characteristic flow of natural waters is one of turbulence, in which irregular motion of subsidiary masses is contained within some relatively simple current. The nature of the turbulence is an expression of various factors including flow velocity, velocity gradients encountered, and the nature of the boundaries of the flowing systems.

Figure may be interpreted in terms of a density current passing through a lake. In this example, the lake proper serves in somewhat the same sense as a stream bank in offering resistance to moving masses. The velocity and magnitude of the eddy systems operating at the upper and lower zones of the density current, for example, are functions of velocity factors of the lake proper and the current, and boundary characteristics including temperature and chemical composition.

It appears, then, that through eddy effects, mass and molecules can be exchanged from one layer to another. One effect of eddies is that by which masses leaving one layer carry with them momentum from that layer and acquire the momentum of the new layer before returning to the original one. This is a simplified definition of *eddy viscosity*. In practice, a *coefficient of eddy viscosity* may be calculated from:

$$r_s = -A\frac{d\bar{u}}{dn}$$

In which rs is the Reynolds stress, or shearing stress per unit surface area; is the shear of the velocities as measured;

$$\frac{d\bar{u}}{dn}$$

And A (from "Austausch") now becomes the coefficient of eddy viscosity, that is, the relationship between mass and transverse distance through which a liquid flows per unit time in a path corresponding to that of the main direction of flow. The transfer of motion through eddy viscosity increases with turbulence, thereby giving rise to the fact that eddy viscosity is considerably greater than molecular viscosity, often on the order of 1000 to 1 million times. In natural waters eddy viscosity is a primary force in reducing the settling rate of sediments, and in maintaining the position of microscopic organisms.

In the same sense that eddy viscosity relates the transfer of mass and momentum from one water stratum to another, *eddy conductivity* describes the exchange of heat across surfaces. This parameter takes into account the specific heat of the fluid and the heat gradients encountered between masses. This heat transfer is then proportional to the mass transfer described in the formula for eddy viscosity. Important here is the idea that conduction may be lateral as well as vertical. Eddy conductivity is fundamental in the heating processes of lakes, described in the preceding chapter.

The transfer of dissolved substances, so very important in lake metabolism and maintenance of living organisms, is another function of mass exchange and is termed *eddy diffusivity*. An important concept of eddy effects is that of the transfer of more than one substance, or the *principle of common transport*. This principle holds that a motion that transfers one environmental factor may, simultaneously, carry another property.

For example, while dissolved salts are being transported downward, dissolved gases, or heat, may be carried upward in the same eddy system. Some idea of the rate at which substances may be distributed through this process is gained from the fact that summertime eddy diffusivity in the western part of Lake Erie is near 25 cm^2/sec, or probably some 104 times that in water in a quiet bottle. The basic concern with eddy effects for our purposes relates to the more general ways in which the processes contribute to lake dynamics.

PERIODIC CURRENT SYSTEMS

To this category belong water movements exhibiting some form of rhythm or periodicity. Two major systems are to be considered; these are: traveling, *surface waves* and standing waves, or *seiches*. As has been shown for nonperiodic currents in the preceding section, external forces, mainly wind and other meteorological factors (atmospheric pressure), are responsible for periodic currents.

SURFACE WAVES

Lake surfaces are never completely smooth; slight irregularities and turbulences are present. Friction occurs between these surfaces, and wind blowing across the lake results in movement of the water. This motion imparts the first phase of wave formation. Subsequently, a crest is built, and increased wind velocity tends to build up the wave; eventually, an eddy is produced by wind on the trailing slope of the wave, giving a certain thrust to its back, thereby accelerating the motion of particles in their orbits within the wave.

There is no essential horizontal movement to waves in open water. A particle of water moves in a vertical circular pattern, or orbit, returning to its original position as the wave rises and falls. The diameter of the orbits and, correspondingly, the velocities, decrease rapidly with depth. It has been calculated that the diameter of orbits at a depth equal to onehalf the wave length is on the order of 1/500 that of surface orbits, or quite imperceptible.

Not until the wind develops to such a force as to produce a tumbling crest, or whitecap, is there any forward motion of the water. A similar effect is exerted by the bottom in shallow zones where friction directed upon a wave causes the mass to pitch forward, producing *surf*. These breaking waves are called *waves of translation*; open-water swells are termed *waves of oscillation*.

Both velocity of wind and the time that it has been exerting force upon a wave determine the size of waves. In order to build waves of any great amplitude a considerable amount of time is required to transfer the wind energy into the kinetic and potential energy of a wave. Similarly, space is required, for a wave must have area in which to move before the wind in order to gain amplitude. The relationship between wave height and fetch during strong winds is expressed in:

$$b_w = 0.105\sqrt{x}$$

In which x is the fetch, or downwind distance (in centimeters) from shore to location of the wave in question. Although waves and wave action are of especial interest to physical limnologists, our interest in terms of the over-all ecology of lakes is mainly in the action of waves in circulation of vital materials in lake waters, and the stirring effects of breaking waves in shallow-water zones. From a broader point of view, wave action is most instrumental in sedimentation and erosional processes which serve to modify the morphology of the lake basin and littoral areas.

SEICHES

A seiche is a form of periodic current system, described as a standing wave, in which some stratum of the water in a basin oscillates about one or more nodes. This rocking motion can be seen in a bowl of thin soup passed by a shaky waiter, the seiche being set up by movement of the basin. By blowing upon the soup near one side of the bowl one can create a seiche due to pressure changes, although the basin be fixed. Seiches in lakes have been recognized for many years, and since Forel's pioneering work, beginning about 1870, much attention has been given to this intriguing phenomenon. The name, attributed to de Duillier, writing in 1730, is derived from the Latin *siccus* through the French meaning dry, and referring to the exposure of low shorelines left dry as the lake water recedes during an oscillation.

We have seen previously that sustained blowing of wind across the surface of a lake will cause water to pile up in the down-wind region of the basin. If the wind suddenly subsides a current will naturally flow toward the area of lowered surface. The water mass does not come immediately to equilibrium. As a result of the energy of motion imparted by the gradient current, an oscillation occurs about a stationary node. This oscillation, a *surface seiche,* continues until damped by contact with the basin proper or by meteorological forces.

The origin of surface seiches may be found in a number of natural mechanisms. From the considerable study of lake oscillations, however, it appears that surface depressions resulting from sustained winds, and from local sudden changes in atmospheric pressure, are the primary agents.

There are cases in which local rain showers set up a seiche due to the pressure of the impact of the falling rain. Sudden inflow from a contributing stream has been suggested as the origin of a seiche in Loch Earn, Scotland. Earthquakes have been named also as possible causes of seiches.

Since these standing waves operate according to the laws of oscillating systems, the formation of harmonics is a general characteristic. In addition to the simple, uninodal seiche, binodal and trinodal seiches have been reported commonly. In fact there is at least one record in Loch Earn of a seiche exhibiting a nodality of the sixteenth. Normally, the periods and locations of the nodes are determined by morphological features of the basin, i.e., depth, diameter, and form. The amplitude of the seiche depends upon the source and the intensity of energy giving rise to the oscillation and to the form of the basin.

Although seiches occur in all enclosed or partially enclosed bodies of water, in small basins the effect is detectable only by sensitive recording devices. In larger lakes the oscillations may be quite conspicuous.

Because of the narrow form of Lake Geneva near the city of Geneva, waterlevel fluctuations due to seiches often leave the shoreline dry. Here, in 1841, a seiche with an amplitude of 1.87 ni was recorded. In the same lake in 1891, Forel observed a seiche which lasted for 7 days, 17 hr.

This standing wave underwent 150 oscillations, the amplitude dropping from a maximum of 20 to 7 cm; the period of oscillations was of the order of 73 min. The Great Lakes exhibit oscillations, the periods of longitudinal seiches in Lake Eric are near 790 rain, and for Lake Huron 289 min. Transverse sciches are also known and serve to complicate measurement of lake activities.

Amplitude and period of seiches are determined in practice by the use of a float contained in a protective housing. The float is connected to a counterbalanced pointer over an appropriate scale. Termed a *limnometer*, the device may employ a tracing stylus to give a graphic recording or limnograph. The periods of seiches can be calculated from various, and often complicated, formulas. For a uninodal seiche in a rectangular basin of regular bottom the period t may be calculated from:

$$t = 2L\sqrt{gh}$$

In which *L* is the length of the basin, *b* is the water depth, and *g* is gravity acceleration. Bearing in mind that a seiche is water in motion, it becomes apparent that turbulence will enter into the system as soon as oscillation begins. The effect of turbulence will be to damp out the oscillation as well as to increase the period.

Thus far, our discussion has dealt with the *surface seiche*. Since about the turn of the century, however, it has been known that internal, periodic current systems could be present in lakes under certain conditions. Within a stratified lake of two or more layers of differing densities the strata may oscillate with respect to each other without being apparent at the lake surface. Recent compilations of data relating to *internal seiches* indicate that both the ranges and the periods of such seiches are significantly greater than those of surface seiches.

During summer stratification the thermocline may, as a matter of character, oscillate between the lighter layer of the epilimnion and the denser hypolimnion. An oscillation of this nature can be clearly and relatively easily detected through temperature recordings at various depths.

Data derived from laboratory and field studies of displacement of lake layers of equal temperature (isotherms) have led to the general acceptance of two major forms of internal seiches. One type, the thermocline seiche, results from the movement of water masses between the epilimnion and the most dense zone of the metalimnion. Another system, the *hypolinmion seiche,* moves along the boundary between the metalimnion and the hypolimnion.

The time of one complete oscillation (period) is usually greater in the case of the hypolimnion seiche than for the thermocline seiche, both, however, being more or less dependent on the length of the lake. The differences in periods are due to unequal densities of the two strata. In Lake Geneva (about 37 km in length), the period of an internal seiche is about four days.

The mechanisms responsible for setting up interval waves are the same as for the surface seiche. Wind-driven waters of the lake surface pile up at the windward end of the basin, causing the deeper waters at the leeward end to compensate by being pushed up. If the lake is stratified, however, the resultant effects of the displacement are different.

Because of greater density difference between air and water than between cool and warm water, slight depression of the water surface by air in motion results in a proportionately greater displacement of the lower water stratum toward the surface in the windward end of the lake. As the wind diminishes, the isothermal strata oscillate about a node near the centre of the lake. The metalimnion serves as a sort of "cushion layer" on which the epilimnion moves over the hvpolimnion.

The problem of internal sciches is not a simple one, either in terms of seiche motions and dynamics or definition. All internal oscillations in lakes may not be "true" seiches, that is, motion resulting from free oscillations. Forced oscillations, or those that are sustained by direct wind action, may be present and sometimes mixed with true seiches. In very shallow lakes forced oscillations may damp the effects of seiches, making difficult a distinction between the two.

Internal seiches are of considerable importance in the total economy of the lake, probably more so than surface sciches. In Lake Victoria an internal seiche with a period of about 30 days brings about the displacement of water into remote parts of the lake basin. The resultant turbulence mixes lake water over the bottom materials and transports dissolved and suspended materials into the shallow zones.

In addition to this more or less horizontal transport, internal seiches serve to distribute heat and nutrients vertically within the lake. One particularly interesting problem is created by the action of internal seiches; this is the determination of the position of the thermocline through temperature measurements. The variations in position of the thermocline when in oscillation. It has become increasingly apparent that the thermocline is not a stable stratum, but rather is subject to much shifting and modification during the season.

True, or lunar tides, have been reported for large lakes, although the amplitude of lake tides would naturally be small. In Lake Superior the range is about 1 in. (3 cm). The difficulty in observing and measuring lake tides arises by seiche complication; a seiche in phase with lunar time (12.5 hr) would magnify the tide range, or if out of phase, would inhibit tidal action.

STREAM CURRENT SYSTEMS

The major processes and properties of flowing waters have been described the importance of such processes in the formation, maintenance, and alteration of the stream or channel. It is suggested that so much of that chapter as pertains to the dynamics of stream flow be reviewed at this time, for those same processes and attributes contribute to the several generalities to be considered in this section.

We have seen that wind is the major force in creating current systems in lakes. In streams the important factor in producing water movement is gravity. So long as a gradient on the earth's surface exists, water flows in response to gravity, to "seek its lowest level" following the route of least resistance. Because the route is seldom straight, the gradient rarely uniform, and the sides and bottom of the channel not often smooth for any distance, varying velocities and turbulences result. Bearing in mind the importance of these factors and activities in channel structure, let us now consider them with reference to the stream as a living-space.

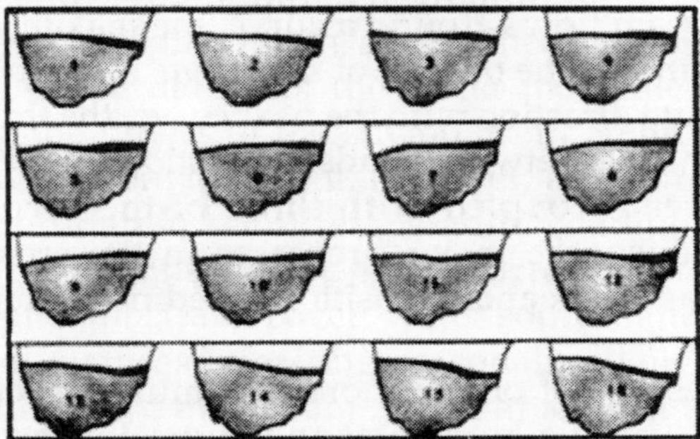

Fig. An Internal Seiche.

A conspicuous characteristic of stream flow is transport of dissolved substances, suspended materials, and living plants and animals. This ability to move matter also contributes greatly to the unstable nature of bottom types and shore zones, mainly through shifting of bottom materials and through sedimentation. Although these processes and features could result from nonturbulent flow, they are usually of greater magnitude in turbulent waters.

Turbulence in streams is not uniform throughout a cross-sectional area of the stream. Maximum turbulence occurs at some region in the cross section and diminishes from that zone toward the surface or the bottom. Zones of maximum turbulence are usually located on either side of a similarly restricted *thread of maximum velocity*. The number of high velocity threads and turbulent zones apparently increases with channel width, depth, and nature of the bottom. The result is decrease in stream stability, making for a more rigorous environment for plants and animals. Eventually, however, uniform discharge may contribute toward a balance between the bottom materials and turbulence and the stream bed becomes more tenable for organisms.

Maximum velocity threads and associated turbulent zones do not maintain a constant position in the stream cross section, except perhaps in straight streams occupying uniform channel shape. Most streams curve and bend throughout their courses, and the velocity threads follow courses that curve even more than do the streams.

As a stream flows around a curve, the maximum velocity thread is found on the outside of the curve; if the stream bends in the opposite direction with the next curve, the thread moves across the channel between bends. The actions of the maximum velocity thread, coupled with those of the accompanying turbulent zones, are major processes in the erosion of the outside stream bank and deposition of sediments on the inside of the curve.

The presence of one or more maximum velocity threads gives rise to varying horizontal and vertical velocities within the steam proper. Such are in evidence as we view the body of a stream and observe the morphology of the stream bed. We are impressed by the presence of a large sand bar separated by a deep channel from, perhaps, a gravel bed; farther downstream the bottom may be strewn with large rocks or be quite muddy.

These features attest to the ever-changing nature of the stream resulting from shifting of the velocity threads and the turbulent zones, and further, to the carrying capacity of the current itself as a function of velocity. Obviously, the relationships are not nearly so precise as indicated here. The velocity varies with the depth and with obstructions such as boulders. In other words, an organism attached to the bottom or to the downstream side of a large stone may not actually be inhabiting a current velocity indicated by the general nature of the stream bed. Swirls and eddies of widely ranging proportions constitute another form of current systems in streams. These systems serve to mix stream waters and to effect deposition of organisms, organic debris, and sediments. Many forms of algae are adapted to eddy dispersal of vegetative structures by taking advantage of shallow, well-lighted swirls for rapid reproduction and subsequent distribution as the eddy contributes to the main current.

Density currents due to differences in temperature, chemical composition, or to silt load occur in streams, especially large streams with smooth bottoms. Spring-fed tributaries are often cooler than the main stream, and as the colder waters meet with the main stream a stratification results. In the lower reaches of a stream or, more properly, in the estuary, salinity currents move with the tides underneath the fresh water. Near the mouth of the Escambia River, in Pensacola Bay, Florida, salinity measurements made in the autumn indicated a surface salinity of 4.5 parts per thousand while at the bottom (about 4.5 m) the salinity was 24.4 ‰. This sharp salinity stratification was also reflected in the freshwater and marine fish distribution.

Tides exert influence on river flow, often well beyond the brackishwater zone. This effect serves to increase turbulence in the lower stream course, and also to bring about slight periodic flooding of the valley. This latter process increases the exchange between stream and land. The Paumunkey River flows on the Piedmont plateau and coastal plain of Virginia, and salinity effects are felt some 12.8 km from the coast. Tidal influences, however, reach about 64 km upstream.

From the consideration of fluvial processes, and the brief review and synthesis here, it is immediately apparent that current systems in the various streams regions as well as in the stream as a whole are extremely complex. Actually, these physical aspects of stream dynamics have received little extensive study in the field under natural conditions. A greater amount of study has been carried on in hydraulics laboratories, and considerable theoretical work has been accomplished. For further information, textbooks in hydraulics, and professional publications such as those of the United States Geological Survey, should be consulted.

CURRENTS IN ESTUARIES

The three most important factors operating to produce currents in estuaries are oceanic tides, stream flow, and wind. The interactions of these forces, particularly the somewhat antagonistic processes of oscillating tides (their vertical ranges and lengthwise flow) and unidirectional stream flow (its velocity and volume), serve to make the estuary a restless and complex system of water movements. Additionally, the morphology of the basin of the estuary and the channel of the stream modify and determine the stream and tidal dynamics. It should be borne in mind that these forces are not regular and constant.

Stream flow varies seasonally with rainfall, while tide height and movement are correlated with lunar effects and wind. Winds may serve to increase either tidal movement or stream flow, depending upon the direction and intensity of the moving air mass.

Within most estuaries there exists a relatively regular and uniform rate of water transport. This is to say that the volume of estuarine water discharged to the sea through the mouth of the estuary essentially compensates for the amount of fresh water introduced by the stream into the upper region of the estuary. In an estuary where the incoming fresh water is mixed with the estuarine water, i.e., where no stratification of the two exists, considerable time may be required for the discharge of a particular mass of the fresh water. This time interval is called the *flushing time.*

It is an average figure used to describe the period during which a quantity of fresh water derived from stream or seepage remains in the estuary. A simplified formula for determining flushing time is:

$$t_F = \left(\frac{S_8 - S_f}{S_8} V_f \right) \frac{1}{v_D}$$

in which S_s and S_f represent the salinity of sea water at the entrance to the estuary and the salinity of the estuarine water, respectively; V_f is the volume of water in the estuary; V_D is the daily volume movement into and out of the estuary. Flushing time in Great South Bay, Long Island, through Fire Island inlet is of the order of 48 days, or approximately 96 tidal cycles.

Obviously a parcel of fresh water entering the estuary near the mouth will move out faster than one in the upper reaches. The main point is that, on the average, the linear movement of dissolved substances and suspended materials, including living plankton, is a slow process. In many instances the time is sufficient for biological events, important in maintaining the integrity of the estuarine community, to occur.

In a "mixing estuary" the stream flow is held back and often reversed by a strong flood tide. Because of the forces of the two opposing current systems, the reversed flow up the stream is normally slow, ceasing completely as the peak of the flood is reached.

Upon the ebb, the outward tidal flow is strengthened by stream flow, the velocity decreasing after the mid-time of the ebb and the movement coming to rest at dead low water. As we have seen for streams, the cross-sectional velocity of an estuary is not constant; a thread of maximum velocity exists. In estuaries, this thread is usually near the surface in the middle of the water course, the precise position depending, however, upon the nature of the shores and bottom.

In many estuaries the typical current pattern is one wherein the lighter, fresh water flows seaward over the upstream movement of denser saline waters (denser by virtue of having a greater concentration of dissolved salts). Under these conditions, a *vertical salinity gradient* exists, and the estuary is said to show stratification. Studies on Chesapeake Bay have indicated that there is a net horizontal seaward flow in the upper layer, and in the lower layer a net horizontal headward flow. The boundary between the two is not level, but slopes toward the right (looking downstream).

In the Tamar Estuary of England, it has been found that at some distance upstream both currents may be seen at the surface. During times of considerable fresh-water flow, the more saline waters of the incoming tide move to one side of the basin. Where stratification and net seaward-headward movements exist, it is possible for certain nutrients and organisms to be transported upstream in the bottom currents, while other materials may be carried to the sea in the upper currents of less saline water.

It has been suggested that oyster larvae are moved upstream by lying on the bottom during an ebb tide and then rising into each successive flood tide to be more or less passively transported for the duration of the tide.

In the preceding paragraphs we have seen some of the effects of oceanic tides on the dynamics of the estuary of a stream. In a great number of instances tidal effects are not restricted to the brackish zone, but occur throughout a considerable length of the stream-estuary-sea continuum. Indeed, in the low-lying course of the Amazon, tides serve to bring about measurable variations in stream depth some 960 km above the salt-water zone. The result is the entrance of tides into the stream before the preceding tide has ebbed. In the Amazon there are said to be eight tides along the stream course at one time.

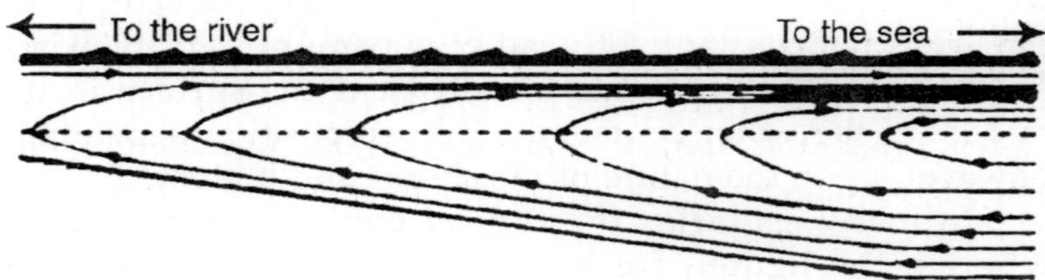

Fig. Schematic Presentation of Current Flow Pattern Along the Central Axis

The morphology of the stream channel and estuary basin regulates, to a great extent, the velocity and magnitude of tidal currents. In a wide, unrestricted estuary, currents are typically slow, without significant turbulence.

In a narrow, deep basin, rapid, turbulent currents generally occur. If the mouth of the estuary is restricted by depositional features or land closures, the incoming tide may be held back until it suddenly breaks forth into the basin as a tidal wave, or *bore*. In the Severn Estuary, England, the tidal range may approach 15 m. A bore of 3 m (sometimes reaching 7.5 m) has been reported in the Tsientang River of China. Such currents exert profound effects on the nature of the substrate, turbidity, and biota of the estuary.

Seiches, similar to those described in lakes, may occur in coastal bays and estuaries. The period of these long stationary current systems is determined by the depth and horizontal parameters of the basin. Oscillations are apparently common in many coastal waters, the effects often obscured, however, by tides.

A trinodal oscillation has been recognized in San Francisco Bay, and the great water-level fluctuations of the Bay of Fundy result from the synchronization of tide and seiche, the latter having an amplitude of some 15 m. The causes of seiches in bays and estuaries are not completely known. It is probable that forces such as variations in atmospheric pressure and wind can supply sufficient energy for the development of standing waves in coastal bodies. As in lakes, friction operates to damp out the oscillation.

The importance of waves and wave action in the estuary is primarily a function of the morphology of the estuary basin. Wherever the mouth of the estuary is restricted and surface area of the body of water slight, there is little opportunity for the development of waves of any appreciable magnitude. Consequently, the role of wave action in watermixing and erosion is negligible. On the other hand, broad estuaries having wide mouths are subject to receiving the full effect of oceanic waves in addition to local disturbances.

Generally, however, the work of moderate Winds are also influential agents in producing certain currents in estuaries. We have already considered the possible role of wind in the development of seiches. Wind contributes greatly to the unusually high waters moved into coastal features during hurricanes and other storms. Depending upon duration and intensity, wind storms may temporarily disrupt the normal circulation patterns in estuaries. Similarly, as the wind is directed, the rate of ebb and flood of tides may be influenced.

GASES IN LAKE, STREAM, AND ESTUARINE WATERS

Many extraordinary properties of water, probably none contribute more to maintaining life in aquatic communities than the capacity of water to hold substances in solution and the ability to enter into chemical reactions. These characteristics result largely from the rather weak nature of the hydrogen bonds linking water molecules.

A relatively small amount of energy is required to separate or to unite the molecules. Therefore, natural waters, including rain, always contain some quantities of dissolved gases and inorganic and organic substances. To repeat an oft-stated truism, "pure water does not occur in nature." On this fact rests the physical-chemicalbiological structure of natural bodies of water that maintains ecosystem continuity as well as evolutionary mechanisms at the organismal level.

The naturally occurring substances that contribute to the "impurity" of waters are derived from various sources, are present in varying states and quantities, undergo certain transformations, and contribute differently to the over-all metabolism of the water community. Certain gases in proper proportion are essential to respiration and photosynthesis in aquatic situations; other gases are lethal to life.

Certain dissolved mineral salts serve as nutrients for free-floating plants; other salts may limit life through osmotic effects. Dissolved organic materials are also present in natural waters, but their reactions are not well known. The nature, quantity, and transformation of substances dissolved in natural bodies of water are generally indicative of the origin of the basin or channel, the climatic regime of the area, and the composition of the substrate in the drainage system.

In this and the following chapter we propose to consider certain properties and reactions of some of the more conspicuous substances contributing to the chemical nature of natural waters as an environment. Although our approach is that of naming the materials and considering them in turn, it is most important to keep in mind the fact that the lake or stream is a dynamic system of chemical interdependencies and interrelationships in association with physical and biological features.

DISSOLVED GASES IN LAKES OXYGEN

Of all the chemical substances in natural waters, oxygen is one of the most significant. It is significant both as a regulator of metabolic processes of community and organism, and as an indicator of lake conditions. Hutchinson has succinctly and aptly stated the case for oxygen in saying: "A skillful limnologist can probably learn more about the nature of a lake from a series of oxygen determinations than from any other kind of chemical data. If these oxygen determinations are accompanied by observations on Secchi disk transparency, lake Colour, and some morphometric data, a very great deal is known about the lake."

The oxygen available for metabolic relationships in natural waters is the oxygen held in simple solution. This is not, as some beginning students are given to think, the O in H_2O. The volume of oxygen dissolved in water at any given time is dependent upon:

- The temperature of the water,
- The partial pressure of the gas in the atmosphere in contact with the water,
- The concentration of dissolved salts (salinity) in the water.

The solubility of oxygen in water is increased by lowering the temperature. For example, the solubility increases about 40 per cent as fresh water cools from 25°C to freezing. Place a ruler or other straightedge across the figure to connect a chosen temperature on the uppermost scale with a given per cent saturation on the middle, inclined scale, say 10°C and 100 per cent saturation. Now read the lowermost scale of oxygen concentration in cc/l at the point where the ruler intersects the scale. Interpreted in one fashion we see that 7.7 cc/l of oxygen constitute 100 per cent saturation at 10°C.

Now connect 5° C and 100 per cent saturation and note that under these conditions a greater concentration of oxygen is dissolved at 100 per cent saturation, 8.7 cc/l to be exact. In other words, solubility increases with decrease in temperature. The original purpose of the nomogram is to determine per cent saturation of a given oxygen concentration at various temperatures and altitudes. Altitude must be considered in order to take pressure into account.

At a given temperature the concentration of a saturated solution of a slightly soluble gas that does not unite chemically with the solvent is very nearly directly proportional to the partial pressure of that gas (Henry's Law). The solubility of oxygen also relates to Dalton *"law of partial pressures"* which states that the total pressure of a mixture of gases is equal to the sum of the pressures exerted by each of the component gases. In other words, the solubility of each gas is independent of other gases in the mixture. Under similar conditions of pressure and temperature the solubility of oxygen in water is over twice that of nitrogen and about one third that of carbon dioxide.

The effect of the third factor, the concentration of dissolved salts, upon the solubility of oxygen is that of decreasing the oxygen concentration as salinity increases. At 0°C fresh water at saturation contains slightly over 2 cc/l more oxygen than does average sea water (35 ‰ salinity); at 15°C the difference is about 1.5 cc/l.

Having given attention to the major factors responsible in regulating the solution of oxygen in waters, let us now consider the sources of this gas. Obviously the atmosphere in contact with the lake surface is an unlimited source of oxygen. The volume per cent of oxygen in the atmosphere is calculated to be 20.99, or approximately 210 cc of oxygen per liter of air. This is some 25 times the concentration of oxygen in the same volume of fresh water.

The rate at which atmospheric oxygen passes across the air-water interface and becomes dissolved in the water is dependent upon a number of factors. For one thing, increased wave action or other disturbances at the lake surface results in greater passage of the gas into solution. Secondly, the greater the difference in partial pressure between air and water, the greater the rate of solution.

Thirdly, the less the moisture content of the gas, the more rapid the solution of that gas. Bear in mind, however, that in all of these processes there may be a "two-way" movement, that gas can be lost from the lake to the atmosphere. The *direction* of movement, as well as the *rate*, is determined by the foregoing factors.

Oxygen in natural waters may also be derived from photosynthetic activity. In shallow, nonstratified ponds lacking significant wave action, oxygen may be derived mainly as a by-product of carbohydrate synthesis by rooted plants and by phytoplankton, viz.:

$$6CO_2 + 12H_2O \rightarrow C_6\,H_{12}\,O6 + 6O_2 + 6H_2\,O$$

or mathematically reduced:

$$6CO_2 + 6H_2O \rightarrow C_6\,H_{12}\,O6 + 6O_2$$

Photosynthesis is also a source of dissolved oxygen in large deep lakes. but, in contrast to ponds, this process is restricted to a certain region of the lake. This zone is delimited by the vertical range of transmission of light effective in photosynthesis. This lighted region is called the *euphotic zone.* It extends horizontally from shore to shore and vertically from the surface to a level beyond which photosynthesis-effective light fails to penetrate.

As we have seen, turbidity, Colour, and the absorptive effect of water itself, serve to quench light; thus they essentially determine the euphotic zone. In the shallow shore region (the *littoral*), submerged rooted plants along, with phytoplankton contribute oxygen to the lake. In the euphotic zone of the open lake (the *limnetic*), phytoplankton contributes the autochthonous oxygen. In most temperate lakes during summer thermal stratification the euphotic zone corresponds closely with the epilimnion.

This accounts mainly for the low-oxygen conditions in the hypolimnion as briefly considered previously. Within the near-surface region of maximum photosynthesis and oxygen gain from the atmosphere, the water often becomes supersaturated with oxygen at the height of diurnal production cycles. Maximum oxygen production usually occurs in the afternoon on clear days, the minimum immediately after dawn. Cyclic daily fluctuations in 'oxygen concentration, called the *oxygen pulse,* have been observed when conditions are proper.

The lower limit of the cuphotic zone is marked by a level at which organic respiration and decomposition consume oxygen at a rate equal to that at which oxygen is produced over a 24-hr period.

This is the *compensation level;* it will enter significantly into our discussion of productivity. Below the compensation level in our typical lake lies the *aphotic zone,* a zone in which there is insufficient light to maintain oxygen production at compensation.

This is the region of low oxygen content, often depleted. It frequently corresponds with the hypolimnion.

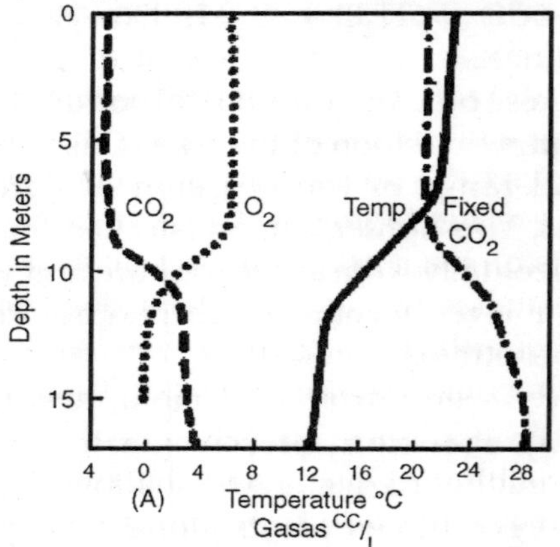

A vertical oxygen distribution pattern such as we have been considering is characteristic of moderately productive lakes of small size.

These lakes, termed *eutrophic* ("rich food"), typically exhibit a hypolimnetic loss of oxygen, the curve of the oxygen distribution dropping sharply through the metalimnion and described as *clinograde.* It seems generally agreed that temperature and oxidation of organic materials are the major factors in creating the oxygen deficiency in the hypolimnion. The rate and magnitude of this oxidative breakdown are dependent upon the volume of organic substance supplied to the process.

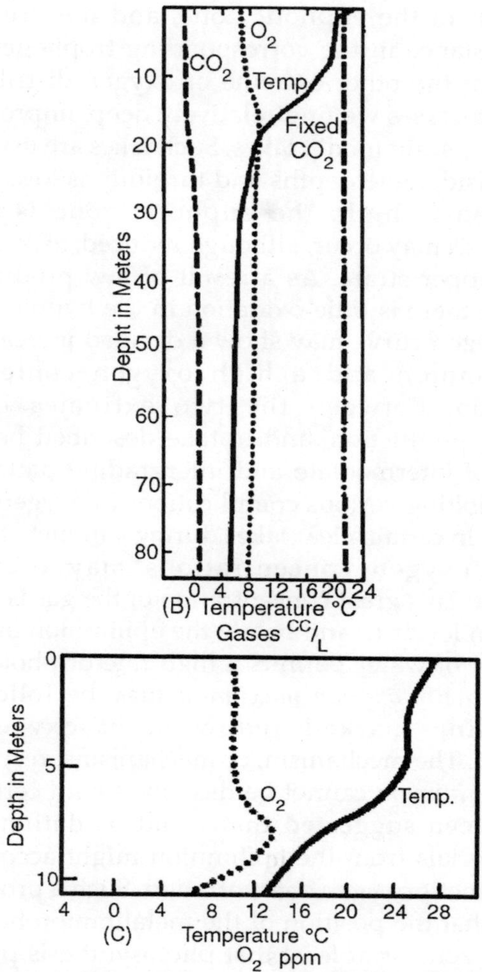

Fig. Various Forms of Oxygen Curves in Lakes. (A) clinograde curve of vertical oxygen; (B) orthograde curve (C) orthograde curve with decrease in oxygen near the bottom.

This substance for the most part is produced in the upper lighted zone, which, in relation the organic synthesis, is called the *trophogenic region*. From here the materials settle into the zone of decomposition, or *tropholytic zone*. It is now apparent that correlations exist among epilimnetic circulation, light transmission in the euphotic zone, and the production of organic substance in the corresponding trophogenic region. At somewhat the other extreme of oxygen distribution and biological processes we find relatively deep, unproductive, or oligotrophic ("scant food") lakes. Such lakes are usually rather clear for considerable depths and turbidity is low. Since light transmission is high, the euphotic zone is deep and photosynthesis may occur, although reduced, at some distance below the upper strata. As a result of low productivity per unit volume there is little oxidation in the hypolimnion.

The oxygen curve may show a decided increase through the metalimnion and a high oxygen content in the hypolimnion. Between the two extremes in oxygen distribution, production, and uptake described here exists a vast range of intermediate and intergrading patterns, many of them exhibiting various complications in oxygen states and metabolism. In certain clear lakes during summer stratification very high oxygen concentrations may occur in the metalimnion. This great concentration of the gas is apparently derived from levels in and below the epilimnion in which the transparency of water permits a high rate of photosynthesis. The *metalimnetic oxygen maximum* may be followed by a corresponding marked reduction of oxygen in the metalimnion. The mechanism, or mechanisms, responsible for this *oxygen minimum* cannot be described with certainty.

It has been suggested that rapid oxidation of slowly settling materials from the epilimnion might account for the low metalimnetic oxygen concentration. Such a process would necessitate that the position of the metalimnion be below the trophogenic zone or at least that photosynthesis proceed at a reduced rate.

Note that underneath the ice in March the concentration of dissolved oxygen is 14 ppm. This saturation was acquired during autumn circulation accompanied by falling temperatures, the latter, of course, increasing the solubility of the gas. In winter some of the oxygen is consumed by organismal respiration and by organic decomposition. In deep lakes oxygen consumption is relatively small, but in shallow bodies of water the oxygen may become depleted, thereby resulting in mortality of animals. Vernal circulation redistributes the oxygen as shown for late April and early May. With the development of summer stagnation the hypolimnion becomes quite poor, often completely lacking in oxygen.

In meromictic lakes the oxygen distribution reflects the essential two-region structure of the water mass. Depending upon the circulation pattern above the chemocline, the oxygen content may take the form of a clinograde curve. The waters of the monimolimnion are usually anaerobic. Lakes of the tropical and subtropical regions exhibit widely diverse oxygen relationships.

Polymictic lakes by virtue of continuous circulation, exhibit a generally high, uniformly distributed, oxygen content throughout the year. Oligomictic waters, in which circulation seldom occurs, are typically characterized by a condition of hypolimnetic oxygen depletion. In shallow lakes of the colder regions, for example Alaska, oxygen at near saturation appears to be uniformly distributed by circulation from surface to bottom. The transport of dissolved oxygen throughout a lake is accomplished primarily by currents set up within the lake and by eddy conduction. Lake currents, as we have previously seen, may be set into operation by wind action on the surface and by density differences between layers in the lake. Circulation in the epilimnion during summer distributes oxygen within that zone.

Eddy systems thrown up along the interfaces between thermal layers or along stream density currents moving through the lake serve to move substances across boundaries, usually at right angles to the current. During seasonal overturn periods the sinking of upper, dense water masses also delivers oxygen to the underlying regions, thereby accounting for the homogeneous oxygen content during those times of uniform density. Simple diffusion of oxygen molecules plays a minor role, indeed, in oxygenation of natural waters.

The process is very slow. As a matter of fact, it has been shown that to raise the oxygen concentration 0.4 mg/l at a depth of 10 m by diffusion from the surface would require over 600 years. With respect to the utilization and ultimate fate of dissolved oxygen in lakes, many aspects have already been considered. In summary, we might say that the major processes acting to consume oxygen are animal and plant respiration, and organic decomposition.

During long periods of cloudiness or where waters may otherwise be shaded, for example underneath broad floating mats of vegetation, organismal respiration and decomposition may rapidly take up oxygen. Where decay bacteria occur in great quantity, as in lakes or streams highly polluted with organic sewage, the water often becomes completely anaerobic.

Although we have been concerned primarily in this section with the maintenance of oxygen in lake waters, the source of the gas, and its distribution and fate, we have had to venture into the subject of productivity. This is a natural consequence because there is an obvious direct relationship between the dissolved oxygen content of natural waters and the amount and rate of energy fixation. Recall that the form of the curve of vertical oxygen distribution indicates a great deal about photosynthesis and decomposition at various levels within a single lake, as well as providing a point of comparison of several lakes.

REDOX POTENTIAL

It has been known in chemistry for many years that oxygen plays an important role in chemical changes in various kinds of solutions. Only within the past 30 or so years, however, has investigation of certain oxygen-related phenomena been transferred from laboratory beaker and battery jar to natural waters of lakes and seas. One aspect of these laboratory and field investigations has been concerned with oxidation and reduction processes.

This line of research has already been fruitful in contributing to our knowledge of chemical limnology and its influence on the activities and distribution of certain organisms. The term *oxidation* is applied to the process in which oxygen is added to a substance, or in which hydrogen is lost from a compound, or in which an element loses electrons.

Conversely, the loss of oxygen, the addition of hydrogen, or the gain of electrons is termed *reduction*. At the elemental level an electron transfer in which *ferrous* iron is oxidized to ferric iron may be stated:

$$Fe^{++} \rightarrow Fe^{+++} + e$$

Under certain conditions, the reduced iron may undergo a change back to the oxidized state by a shift in electrons as suggested in the reversibility of the process shown above.

The extent to which a substance can undergo oxidation-reduction processes is dependent upon the concentration of other oxidizing-reducing systems and their products in the solution. Within a given solution, the proportion of oxidized to reduced components of a particular system in relation to other systems constitutes the oxidation-reduction potential, or *redox potential*.

In a solution, for example lake water, a system with a given redox potential undergoes reduction and oxidizes a system of lower redox value. It may also cause a reduction to take place in a system of higher redox potential.

In practice, the redox potential is determined by immersing a nonreactive electrode (bright platinum is often used) in a solution. The second electrode, necessary to complete the circuit, is usually the hydrogen electrode. The presence of the electrode sets up an electron flow, the direction of which depends upon the proportion of oxidized to reduced material. Where an excess of the reduced state (Fe^{++}) exists, electrons flow to the electrode and oxidation in the system results. In a solution containing a surplus of the oxidized state (Fe^{+++}), flow from the electrode contributes electrons and reduction occurs. The excess or deficit of electrons (relative) can be measured in volts on a potentiometer as an electromotive force about the electrode.

This measure gives the intensity of the force (E_h), either positive or negative, and is the redox potential. A positive E_h reading results from a state tending toward oxidation; a negative E_h indicates a system causing reduction. In addition to *intensity*, the oxidationreduction dynamics also has the attribute of *capacity*. The capacity of an oxidation-reduction system refers to the ability of a system to undergo a certain amount of oxidation-reduction transformation without an intensity change.

Oxygen in natural waters produces a redox potential which is influenced considerably by temperature and the hydrogen ion concentration (pH). Out of consideration for pH effects, the potential is measured at the prevailing pH and then referred to pH 7, this measurement being called the *E*7. Correlations between ranges of E_h at pH 7 and oxidationreduction reactions in certain systems have been established as follows:

NO_3- to NO_2-: 0.45 to 0.40 v
NO_2- to NH_3: 0.40 to 0.35 v
Fe^{+++} to Fe^{++}: 0.30 to 0.20 v
$SO_4=$ to S=: 0.10 to 0.06 v

At a temperature of 25°C and pH 7, well-aerated lake waters exhibit a redox potential of about 0.5 v. This potential remains relatively steady as long as the oxygen content is above approximately 1 mg/l. In other words, the redox potential, as such, is affected but little by the oxygen concentration.

Generally speaking, the curve of vertical distribution of the E_l, in lakes follows that of dissolved oxygen and may be a "mirror image" of the ferrous iron. In stratified lakes exhibiting a somewhat orthograde curve of oxygen, the redox potential gives essentially the same pattern. In small lakes in which the oxygen in the hypolimnion is quite low, therefore giving a clinograde oxygen curve, the E_h usually, but not always, appears in a clinograde distribution. Gradations between these forms are common.

Oxidation-reduction activities at the water-mud interface of the lake bottom bear markedly upon the lake chemistry, particularly in the deep hypolimnion, and upon the type of organisms present in the shallow part of the sediments. The E_h of mud exposed to oxygenated water varies near 0.5 v. This potential may extend for several millimeters through an oxidized zone of brownish mud at the surface of the bottom sediments, the *oxidized microzone*.

Accompanying the decrease in hypolimnetic oxygen with summer stagnation is a diminution in the depth of the oxidized microzone. As the $E_{l'}$ of the interface approaches 0.2 v, the oxidized microzone may disappear. Below the microzone, the sediments of deep-water lakes are usually highly reducing in nature, the E_h being near zero volt. An oxidized microzone may be persistent in lakes of size and depth sufficient to exhibit an orthograde oxygen distribution.

The processes responsible for maintaining the integrity of the microzone are not completely understood. It was early suggested that molecular diffusion of oxygen through the mud depended upon the reducing state of the sediment. Turbulence may also be important.

We are agreed that low oxidation-reduction potentials suggest the presence of reducing substances which would, in all probability, utilize such free oxygen as might be brought into the solution. For organisms such as anaerobic bacteria, or other organisms not requiring free oxygen for respiration, a low E_h would pose no problem. Anaerobic bacteria exist where the E_h lies below-0.4 v.

Oxygen-dependent plants and animals would be restricted from such a zone, although they may be found inhabiting places where the Eh is as low as-0.2 v. Zonation of certain insect larvae in lakes has been correlated with E_h. The midge *Calopsectra* (*Tanytarsus*) sp. dominated the insect fauna in bottom muds over which the Eh of the water was 0.4 or above. Another midge, *Tendipes* (*Chironomus*) sp., characterized muds in waters in which the E_h was below 0.3 v.

CARBONDIOXIDE

The very great importance of carbon dioxide as a contributor to the fitness of natural waters as environment derives from essentially three factors. In the first place it serves in a more or less purely chemical sense to "buffer" the environment against rapid shifts in acidity-alkalinity states. In this sense carbon dioxide ameliorates the chemical environment through the ability of the gas to combine with water to form an acid, and to react to give a neutral salt, or a base. We shall give attention to these reactions later. A second contribution of importance by carbon dioxide pertains to regulating biological processes in aquatic communities. Seed germination of some plants, as well as plant growth, is determined by the concentration of carbon dioxide. Various animal processes such as respiration and oxygen transport in blood are related to carbon dioxide. A third and most important contribution by carbon dioxide lies in the fact that it contains carbon.

Carbon is one of the most versatile of all elements, due to the possession of four electrons in the outer ring which give great bonding capacity to form a fabulous number of compounds, many of exceeding complexity. The ability to form many compounds is due largely to the asymmetrical nature of the bonds and to the fact that carbon can form chains of atoms almost without limit.

This latter characteristic distinguishes carbon from other elements and from inorganic substances, although many inorganic compounds, of course, also contain carbon. Carbon dioxide and water supply the carbon, hydrogen, and oxygen which are major components of protoplasm.

The numerous and varied activities of carbon dioxide in the aquatic ecosystem are made possible primarily by the very high solubility of the gas in natural waters. The solubility of carbon dioxide varies inversely with temperature; at temperatures common in nature, carbon dioxide is much more soluble than oxygen in water. At 20°C and atmospheric pressure of 760 mm Hg, water in equilibrium with atmospheric carbon dioxide contains about 0.88 vol of the gas; only 0.031 vol of oxygen are contained in water under similar conditions. Although air contains some 700 times more oxygen than carbon dioxide (by volume), the proportion in water at equilibrium is more nearly equal, about 4 cc carbon dioxide per liter to 6 cc oxygen per liter. If we consider the amount of carbon dioxide "locked up" in various combined forms, the total amount of the compound in water, particularly the oceans, is much greater than in air.

Carbon dioxide in natural waters is derived from a

number of sources. Bacterial decomposition of organic matter in the tropholytic zones, and respiration by animals and plants contribute to the store of carbon dioxide. In the case of plants the greater net contribution is at night, when photosynthesis is not occurring.

Ground waters flowing or seeping into lakes and streams may carry carbon dioxide, the amount being determined by the extent of decomposition in the topsoil and, as we shall presently see, by the chemical nature of the underlying rocks. Within the body of water, certain chemical reactions between acids and various compounds of carbonates release carbon dioxide. Finally, the atmosphere directly furnishes some carbon dioxide to natural waters, and rain, as it falls through the atmosphere, dissolves some of the gas and delivers it to lakes and other waters.

The importance of rain in supplying carbon dioxide to inland waters lies in reactions wherein carbon dioxide is maintained and transported in forms other than as a gas. As rain percolates through soil containing carbon dioxide of decomposition, some of the gas becomes dissolved in rain water. The reaction between CO_2 and H_2O results in the formation of carbonic acid (H_2CO_3). If this weak acid encounters carbonate-holding rocks, limestone ($CaCO_3$) for example, the latter dissolves as calcium bicarbonate, or $Ca(HCO_3)_2$. The solution of $Ca(HCO_3)_2$ remains stable only in the presence of a certain amount of *free* or *equilibrium, carbon dioxide*. Free carbon dioxide represents the CO_2 in $H_2\ CO_3$ plus that in simple solution. The formation of H_2CO_3 and its dissociations are shown in the reactions:

$$CO_2 + H_2O \rightarrow H_2CO_3 \rightarrow H^+ + HCO_3^- \rightarrow H^+ + CO^2_3$$

Note in the above reactions that carbon dioxide is contained in two states not previously encountered, i.e., as bicarbonate and carbonate radicals, $HCO_3^-$and CO_3^{2-}, respectively. This carbon dioxide is called the *combined carbon dioxide*. Solution of $CaCO_3$ is dependent upon the addition of CO_2 in an amount greater than that of free carbon dioxide; this additional CO_2 is known as *aggressive carbon dioxide*. The major points to be gained from these considerations include:

- The relations between rain and soil in supplying compounds containing carbon dioxide to natural waters,
- The chemical reactions by which those compounds are formed,
- The occurrence of carbon dioxide in its three forms: free (CO_2 in solution plus that in H_2CO_3), half bound (HCO_3), and bound (CO_3^{2-}).

The direction of reactions involving the forms of carbon dioxide and the very occurrence of these in natural waters are reciprocally related to acid-base relationships in the medium. When carbon dioxide dissolves, the reaction and end products depend upon the nature of the solvent, particularly with relation to the hydrogen-ion concentration. Under acid conditions, the combination is as shown in Equation above in which H_2CO_3 is formed followed by dissociation to $H^+ + HCO_3^-$ Under highly basic conditions the reaction is:

$$\text{Base OH} + H_2CO_3 = H_2O + \text{base}^+ + HCO_3^-$$

CHEMICAL BUFFERING

The maintenance of near-neutral conditions in mineralized waters is due to *buffering* by chemical systems such as the carbon dioxide-bicarbonate carbonate complex. Other systems may involve magnesium, sodium, or potassium. In other words, as acid conditions arise, the reaction between acid and base from the bound carbonate, for example, brings about an increase in neutral bicarbonate. Continued acidification releases carbon dioxide and carbonic acid from the bicarbonate accompanied by loss of carbon dioxide from the system.

This reaction is the basis for certain limnological techniques for determining the so-called *alkalinity* of natural waters. In these tests a quantity of strong acid is added to water in the presence of a proper indicator. The amount of acid necessary to convert any carbonate or bicarbonate present to free CO_2 is a measure of the HCO_3^- and CO_3^{2-} in solution. In this sense, alkalinity refers to the quality and quantity of compounds which bring about a shift in the pH of a solution toward the alkaline side of the pH range.

Although not always the case, alkalinity usually reflects the activity of calcium carbonate. Therefore, three forms of alkalinity may be recognized: bicarbonate, normal carbonate (*phenolphthalein alkalinity*), and hydroxide.

At the other end, increased alkalinity of the solution brings about a reaction between the base and carbonic acid to hold the departure from neutrality to a smaller value than would otherwise be the case if the buffer system were not present.

DEFINITION OF pH

The pH is the logarithm of the reciprocal of the hydrogen-ion (or more properly, the hydronium-ion) activity. The pH may be expressed mathematically as follows:

$$pH = \log \frac{1}{(H^+)}$$

where (H^+) is the amount of hydrogen ions in a solution in moles per liter. In a liter of pure water there is 0.0000001 of hydrogen mole/ions (and a corresponding quantity of (OH^-)). The pH of pure (neutral) water is, therefore,

$$pH = \log \frac{1}{0.0000001}$$

Increase in the concentration of H^+ ions results in a lower pH value, or conversely, reduction in the H^+ concentration brings about a higher value. From this relationship, a pH scale has been devised. This scale ranges from pH 0, corresponding to a solution with $(H^+) = 1$, through pH 7 (or neutrality) to pH 14, corresponding to a solution with $(H^+) = 10^{-14}$. From pH 0 to pH 7 solutions are *acid*.

From pH 7 to pH 14 reactions are *alkaline*. With respect to the buffering system, we note that only free carbon dioxide is of any import in natural water systems below pH 5, that bicarbonate dominates the range from pH 7 to pH 9, and that the carbonate radical is most important in the range above pH 9.5 or 10.

The pH range of lakes having some degree of flow through the basin is generally from about 6 to 9. In limestone regions the dissolved carbonates may extend the pH range considerably beyond 9. In basins lacking outlets, evaporation may concentrate alkaline substances, resulting in pH readings of over 12. At the other extreme, accumulation of acids such as sulfuric acid in volcanic lakes gives a pH as low as 1.7.

OCCURRENCE OF CARBON DIOXIDE IN LAKES

Having given attention to some of the major aspects of the sources of carbon dioxide in natural waters, as well as to the complex and often critically balanced reactions of the forms of the compound, let us now turn our attention to general considerations of the compound in lakes.

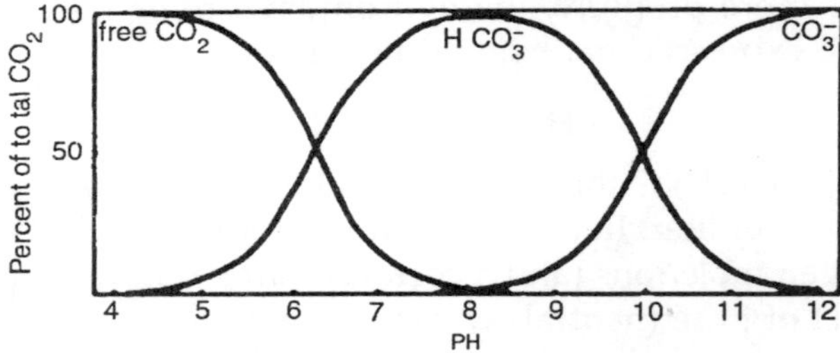

Fig: The Relationship of Hydrogen Ion Concentration to the Percentage of Total Carbon Dioxide in Each of Its Forms in Water.

From what we have seen of the close relationship between the chemical nature of the drainage basin substrate and the chemistry of waters of the basin, we should expect wide regional variation in carbon dioxide content of lakes. Newly formed lakes in regions of weakly soluble rocks typically contain little carbon dioxide in any form.

In these situations the low quantity of carbonates in the substrate results in little of the bound or half-bound forms of carbon dioxide. Similarly, the paucity of soluble minerals as nutrients in biological processes inhibits the development of large biotic populations that would contribute carbon dioxide through respiration, and, of course, decomposition is reduced. Lakes of this type are usually slightly acid, the pH ranging near 6. Lakes of higher acidity (pH 4 to 6) are common in regions of lowlands and bogs. In these waters the free carbon dioxide content is usually quite high, ranging to nearly 200 ppm.

As would be expected in view of the low pH, the concentration of bound carbon dioxide as carbonate is low, usually less than 9 or 10 ppm. Lakes having these characteristics are termed *soft-water lakes*. A large number of lakes fall into a class in which the pH is circumneutral, and which may be called *medium-water lakes*. Free gaseous carbon dioxide in medium-water lakes varies widely, frequently showing supersaturation relative to the partial pressure of the gas in the atmosphere.

These lakes may contain bound carbon dioxide up to 30 or 35 ppm. In regions where the substrate contains easily dissolved minerals, *hard-water lakes* occur. These lakes are characterized by negative values for free carbon dioxide due to of bicarbonates at a greater rate than carbonates are precipitated, and by pH values ranging from about 8.5 upwards. Bound carbon dioxide content amounts to over 35 or 40 ppm, often reaching 200 ppm or more.

SEASONAL CYCLE OF CARBON DIOXIDE AND pH

During winter in the colder regions ice cover forms over the lake, thereby inhibiting exchange of materials across the air-water interface. Thus the lake becomes essentially "sealed in." Oxidation of organic substances, particularly in the depths, consumes oxygen and increases the carbon dioxide content. This often results in the building up of a gradient of metabolic substances from surface to bottom, especially if there is some transmission of light through the ice permitting photosynthesis in the upper, unfrozen waters.

Under such conditions the waters immediately below the ice may contain a considerable quantity of oxygen, sometimes reaching near saturation, and little or no free carbon dioxide. The pH in this zone may show slightly alkaline conditions with low bicarbonate concentration and absence of carbonates. Remember that we should expect to find carbonate only in the absence of free carbon dioxide. In the deep waters, bicarbonates and free carbon dioxide are increased, while oxygen may be nearly or completely depleted; the pH may drop as during summer stratification.

In the warmer regions of the temperate zones, nonfreezing lakes typically contain quantities of carbon dioxide more or less uniformly distributed throughout the waters during winter. This is, of course, due to circulation during the cold season. At this time the phenophthalein alkalinity of surface waters is usually nil, and carbonates are absent.

In holomictic lakes the period of vernal mixing brings about the relatively even distribution of all dissolved materials. During this overturn the pH of the water is uniform from surface to bottom. Free carbon dioxide, together with other gases such as methane and hydrogen sulfide, derived from winter decomposition processes, is lost to the atmosphere. Following the loss of the free form, bound carbon dioxide reappears as carbonate in medium and hard-water lakes as summer stratification begins.

We have already seen that in productive holomictic lakes exhibiting two circulation periods the oxygen concentration diminishes rapidly with depth during summer, giving a clinograde curve. In these lakes the carbon dioxide and carbonate concentration show a general inverse relationship to the oxygen; that is, the concentration of carbon dioxide and bicarbonate increases slightly with depth. On the other hand, an orthograde oxygen distribution is usually acompanied by only slight increase, if any, in carbon dioxide.

During the summer the epilimnion of medium and hard-water lakes is typically devoid of free carbon dioxide, and contains measurable quantities of carbonates. It has been shown, for example, that in Douglas Lake, Michigan, in July, free carbon dioxide may be absent to a depth of about 14 m, and in this same distance carbonates range from about 10 to 8 ppm. pH in the upper waters is slightly over 8. At a depth of 20 m, the free carbon dioxide content reaches 11 ppm at a pH of 7.1. As a point of interest, records over a 30-yr period show that the surface alkalinity due to half-bound carbon dioxide varied only from 110 to 128 ppm.

Autumnal circulation brings about uniform physical and chemical conditions throughout the lake. Vertical gradients in density, temperature, dissolved gases, and solids, which became established during summer stagnation, are broken down. The hypolimnetic gases of decomposition are thrown off at the lake surface. The lake's oxygen content is circulated, and with lowering temperatures of autumn an increased supply is dissolved in the water.

OTHER LAKE GASES

Methane is an organic gas (CH_4) widely called "marsh gas," common in many alkaline lakes, ponds, and swamps, particularly during summer stratification. It is produced by bacterial decomposition of organic substances in the tropholytic zone, primarily in the lake bottom. The process involves a multistage breakdown of complex organic material to compounds of simple molecular structure and then decomposition of these, releasing methane and carbon dioxide. It occurs only under anaerobic conditions. In carbohydrates, for example, cellulose is attacked by bacterial enzymes and hydrolyzed to a simple sugar.

Anaerobic decomposition of the sugars may result in the formation of hydrogen and methane. The time of highest production appears to be during summer stagnation when the bottom muds have become significantly reducing, that is, when the oxidation potential is low. The process can proceed at relatively low, temperatures, about 5°C, because of the tolerance of at least one of the methaneproducing bacteria.

In shallow bodies of water such as ponds and swamps, and in the shore zones of lakes, bubbles containing methane and other gases are often seen rising to the surface and erupting. The formation of bubbles is apparently ssdue to insufficient water in the shallow situations to dissolve the gas as it is formed in bottom mud and debris.

It follows then that the occurrence of bubbles at the surface over deep water would be unlikely. Under winter conditions methane bubbles may be present in the ice cover.

Analysis of the gases in bubbles in a Russian lake (Beloye) revealed a methane proportion of from 74 to about 84 per cent, and 5 to 18 per cent hydrogen. As the bubbles rise to the surface, the hydrogen is lost and the methane volume diminishes to near 24 per cent. In the ascent, nitrogen and a small quantity of oxygen are gained. These transfers of gases into and out of the bubble operate, of course, under the law of partial pressures.

Although the full story of methane and its role is not known, it does appear that some of the gas is oxidized by organisms in oxygenated zones of lakes as the bubbles rise. Here we see yet another phase in lake metabolism serving to decrease the oxygen content of deep waters.

Additional gases occur in natural waters and should be mentioned at this point, even though certain of them occur in small quantities. These will be considered in relation to dissolved solids. As we have already seen, hydrogen is formed as a decomposition product in the anaerobic zones of bodies of water. It acts, in part, to form methane and also occurs free in bubbles.

Free ammonia may, under certain conditions, be present in small quantities in lakes and streams, and elemental nitrogen, derived mainly from the atmosphere, is highly important in lake metabolism; these substances will be taken up with nitrates. Hydrogen sulfide, a decomposition product, is frequently present in the hypolimnion of certain lakes during summer; it will be considered in the following chapter along with sulfur.

DISSOLVED GASES IN STREAMS

The Behaviour and basic relations of gases in streams follow the same fundamental physical and biochemical laws as operate in lentic situations. However, temporal and spatial relationships of dissolved substances, generally, are variously modified and complicated by the very features that characterize streams. In other words, the presence of a current with its inherent turbulence effects, the considerable exchange between stream and surrounding terrain, and the variations in water volume and chemistry associated with climate and drainage basin morphology all serve to make a momentary or long-term picture of stream conditions quite different from a lake.

During our time another factor, an "unnatural" one, has become important in relationships in streams. This is pollution, and although the act of ravaging our waters by dumping wastes into them is not to be condoned, we have learned much about biochemical reactions from pollution studies. This topic will be studied more fully in a later chapter.

OXYGEN IN STREAMS

There are three primary sources of oxygen in stream water, the contributions of each being far from equal and indeed varying greatly with time of day, season, current velocity and stream morphology, temperature, and biological characteristics.

GROUND WATER AND SURFACE RUNOFF

For most streams these sources are relatively insignificant in supplying oxygen. Water issuing from springs, subterranean channels, or from seepages is typically low in dissolved oxygen, often to the point of being anaerobic. Not only do these ground waters fail to provide oxygen to the spring run, but also the run itself dilute the oxygen content of a parent stream at the point of junction of the two streams.

If, however, subsurface waters flow over broken rocks shortly, before reaching the surface, the waters may be near saturation. Similarly, if surface runoff is rapid and vigorous, the water may be high in oxygen content; but sluggish surface drainage seldom contributes great oxygen stores to running waters.

PHOTOSYNTHESIS

In less turbid streams vegetation contributes oxygen to the waters during the day. Rich growths of algae on submerged objects such as rocks or logs, and algae floating free in the water, together with higher plants growing beneath the surface, produce high amounts of oxygen, particularly on cloudless days. During the night and on cloudy days this production may be somewhat balanced by respiratory consumption of the oxygen by plants and animals.

In the shallow headwater reaches of a stream a net production of oxygen by photosynthesis contributes to the downstream content. In the lower, more turbid regions of most streams, however, local photosynthesis probably contributes little to the downstream oxygen content. Since turbidity is to a great extent a function of stream discharge and capacity, we can again recognize the importance of stream channel and water-shed features in the chemistry and biology of streams.

PHYSICAL AERATION

The introduction of a large amout of organic substance such as sewage, or debris from swamp or marsh flooding into a stream may bring about a depression of the dissolved oxygen content below the saturation value. The difference between the actual oxygen content and the amount that could be present at saturation is called the *saturation deficit*. This deficit is incurred, of course, through the uptake of oxygen by aerobic decomposition of the organic materials in the stream. Yet, downstream, barring no further immediate depressions, the stream shows evidence of regaining its earlier level of oxygen concentration.

The oxygen serving to offset the oxidative loss is absorbed from the atmosphere' through reaeration of the stream waters. Reaeration is, therefore, a process by which streams secure oxygen directly from the atmosphere, the gas then entering into the biochemical oxidation reactions in the stream.

Within the stream, distribution of the oxygen derived from the atmosphere is accomplished by turbulent transport. The rate at which reoxygenation of a given parcel of water takes place depends upon a number of factors, including temperature, degree of turbulence, depth of the parcel, magnitude of the saturation deficit and, naturally, the momentary oxygen demand by decomposition processes. The importance of temperature lies in its inverse relationship to oxygen solubility, and to its influence on metabolic demands of organisms. Turbulence is a highly variable factor in the oxygenationdeoxygenation relationship, varying from negligible in quiet pools to highly important in riffles and rapids.How effective is reaeration as a method of oxygenating flowing waters?

Note particularly the relationships between discharge and saturation deficit, and the dissolved oxygen content. The increase in oxygen concentration is greater at the lower discharge than at the higher. The total mass picked up, however, is greater in the higher discharges.Under natural conditions, the waters of streams typically contain a relatively high concentration of oxygen tending toward saturation. However, a number of factors operate to varying extents to reduce the oxygen content and to contribute to the loss of the gas from streams.

- *Turbulence*: We have just witnessed the fact that turbulence plays an important part in the aeration of streams. It follows, therefore, that physical aeration is reduced with decrease in turbulent flow.
- In stream areas below riffles and falls, the water is often saturated with oxygen throughout the day and night. In quiet reaches and in streams of low velocity, the waters may be below saturation during the night.
- *Respiration of Organisms*: Respiratory activities of plants and animals, and oxidation of organic matter utilize dissolved oxygen of streams. The effects of these processes are more conspicuous at night, being masked during the day by photosynthesis.
- *Photosynthesis*: In the more stable zones of streams, submerged plants contribute in a major way to the oxygen content of the water. Consequently, fluctuations in photosynthetic activity will be reflected in the amount of dissolved oxygen present.
- *Temperature*: In streams, as in other waters, solubility of oxygen varies inversely with temperature. Thus, raising of water temperature could result in loss of oxygen from streams.
- *Atmospheric Pressure*: Since the solubility of oxygen bears a direct relationship with atmospheric pressure, reduction of pressure would bring about a decrease in the amount of dissolved oxygen.

- *Inorganic Reactions*: Certain inorganic activities, such as the oxidation of iron, may contribute to the loss of oxygen from polluted streams.
- *Inflow of Tributaries*: The introduction of tributary waters of low oxygen content serves to dilute the concentration of oxygen in the receiving streams. This effect is especially noticeable with the entrance of spring or some seepage waters.

The annual cycle of oxygen of streams is closely correlated with temperature conditions. Studies of large rivers and small streams in warm southern regions of moderate temperature regimes and in northern climates of broad temperature fluctuations have shown that the oxygen content of flowing waters is generally highest in winter and lowest in late summer.

This primary temperatureoxygen relationship may, however, be tempered by a number of factors acting throughout the year. In small, slow streams of northern latitudes, decreased day length and ice and snow cover may serve to inhibit photosynthesis, thereby bringing about a short-term depression of the winter peak of oxygen concentration.

The vernal decline of oxygen content in slow, clear streams may be attributed to the action of spring floods in removing vegetation. Further, decrease in oxygen content toward late summer, in streams generally, may be due to one or more of several factors. Water temperatures reaching their maxima in late summer hold less of the gas in solution; decreased discharge results in diminished physical mixing and reoxygenation; greater decomposition of summerproduced organic material utilizes some of the available oxygen.

On the other hand, a rapid and abundant growth ("bloonm") of phytoplankton often increases the oxygen content during the summer. A daily oxygen rhythm, the diurnal pulse, is largely a reflection of temperature fluctuations and photosynthesis-respiration relationships.

In mountain and hill country the water is typically at between 95 and 105 per cent saturation at all times, so long as there 'is not a dense growth of rooted aquatic plants.

Turbulence is of great importance and easily maintains an average of 100 per cent saturation. In a slow, shallow stream containing a fair abundance of rooted plant growth, daytime photosynthetic production of oxygen exceeds turbulent diffusion into the air and the respiratory consumption of the gas. The result is frequent super saturation with oxygen during the day, followed by a drop to 100 per cent saturation at night. In winter in the temperate zone the time of maximum daily oxygen production corresponds closely with the time of highest temperature. In summer, oxygen production lags behind rising temperature, often by two or three hours, thus imparting high saturation values even after sundown. Depending upon oxygen demand, the minimum level usually occurs prior to the early-morning temperature low. In streams of this naute the oxygen concentration may be acutely affected by sunlight intensity.

The depression in the diurnal curve between about 1300 and 1500 hr is doubtless due to cloudiness shown for the same periods on the light-intensity curve. Such data as these indicate the importance of insolation and photosnythesis in maintainiiig the oxygen content of slow streams. The considerable diurnal oxygen range in such streams, also points up the importance of more than a single oxygen determination in a stream study.

Obviously an average of several measurements would be much more meaningful than one, and a diel series would most satisfactorily describe the oxygen conditions in this stream type. We have already learned that oxygenation through turbulent mixing is a prominent feature of fast waters and even highly turbid streams of low discharge. The diurnal oxygen pulse of such streams as these is usually less pronounced, since mechanical aeration and oxygen consumption are essentially uniform within a given stream segment.

The distribution of dissolved oxygen throughout the linear extent of a stream is subject to many variables, so much so that generalizations relating to the feature are not readily apparent. As we have seen, the upper reaches of streams are usually well oxygenated, the concentration being a function of turbulence and water temperature. Farther along the gradient, the stream becomes slower and rich growths of submerged plants may develop. In this region the oxygen content is influenced less by physical aeration and more by organic oxygen production and respiration.

The maximum-minimum relationships here may be the reverse of those found upstream, in that maximum saturation values will be found in the afternoon and the lowest values in the early morning. In the lowlands, increased turbidity and organic decomposition produce a generally low oxygen content. In clear, slow streams supporting an abundant flora, oxygen may be added to the waters along the stream course, thereby resulting in a net increase in downstream segments.

The introduction of organic pollutants and inflow, from marshes and swamps may serve to decrease the oxygen content at and below the point of inflow of these materials. Thus the amount of dissolved oxygen distributed through the course of a natural stream is strongly determined by channel and flow characteristics and by biochemical interactions and processes.

Although our knowledge of reservoir limnology is meager, we know that the effects of the release of impounded waters into streams below the impoundment are varied and important, and worthy of considerably more research in the future. For example, we know that release of water from a reservoir through deep-water intakes during summer stratification introduces poorly oxygenated water and organic materials from the hypolimnion into the stream below the impoundment. Under extreme conditions this actual dilution of the stream and introduction of oxidizable matter could have deleterious effects on the biota.

On the other hand, draw-down through shallow intakes from the epilimnion may serve to increase the dissolved oxygen content of the downstream waters.

CARBON DIOXIDE AND PH IN STREAMS

In streams the occurrence and abundance of components of the bicarbonate buffer system and the pH condition are determined primarily by current, biological processes, and the chemical nature of the substrate. The role of current is that of ameliorating the chemical climate of the stream. This is accomplished through mixing and moving concentrations of substances, but usually within a relatively restricted segment of the stream.

Over a considerable distance, and depending upon the volume of introduced materials, the chemical composition of stream waters is subject to considerable change. The biological processes acting to influence the nature of the water are those previously considered, i.e., photosynthesis and respiration. In general, the Behaviour of carbon dioxide in streams is similar to that of oxygen, but with inverse properties.

In stream areas of abundant plant growth the concentration of carbon dioxide is minimum during most of the day and maximum in the early morning hours. In sluggish streams a phytoplankton bloom may indeed exhaust the free carbon dioxide supply and obtain further carbon dioxide from bicarbonates. In the lower, sluggish stream course under conditions of high turbidity due to organic suspensoids, high carbon dioxide content occurs in the presence of depressed oxygen concentration, owing to bacterial action.

The chemical composition of mineral-bearing rocks in the stream valley and channel, and also the drainage nature of the valley, may act in a major way to determine the water composition. Under certain conditions these factors may somewhat offset the influences of biological processes. For example, in limestone regions a high concentration of carbon dioxide is not apt to develop in the stream because, as we have already seen, any excess of gas would enter into combination with line in the substrate to give a carbonate.

The direct strong influence of the substrate rests primarily upon the solubility of tile buffer sunstances in rocks. Streams issuing from or flowing over relatively insoluble igneous formations of high silica content are typically soft, because the bicarbonate content is insufficient to buffer pH changes due to accumulation of carbon dioxide.

As a result of a shift of the buffer system toward carbonic

acid, the pH of such streams will be below neutrality. However, atmosphere-water equilibrium essentially adjusts the carbon dioxide content in such a way that the pH value does not go far toward the acid side, usually coming to balance at about pH 6. If, on the other hand, the stream encounters water from bogs, which commonly occur in areas of siliceous formations, the pH may be further lowered.

In many regions, the Appalachian slopes of North America for example, the fate of soft-water streams is to flow eventually over sedimentary formations in the lower reaches. These formations are normally rich in soluble carbonates, and the addition of carbonate ions serves to shift the pH value toward the basic range of the scale, usually to pH 8 or higher. The increase in pH value is accompanied by a marked rise in total alkalinity, and decrease in carbon dioxide.

Highly acid streams occur primarily on low marshy or swampy terrain, on poorly drained sandy "flatwoods," or under special conditions. These streams are usually stained brownish and support a relatively meager biota.

The nature of the acids contained in such streams is not fully known, but frequently considered to be humic, yellow organic, or perhaps sulfuric. In the southern United States, the pH value in sluggish acid streams may range near pH 4, with a concurrent high concentration of free carbon dioxide (25 ppm or more) and low bicarbonate content. Oxygen in these streams is often low, being on the order of 20 to 30 per cent saturation at 17° to 18°C.

Hard-water streams occur commonly as spring runs or surface drainage streams in regions of soluble basic geologic formations. Calcium and magnesium carbonate, often occurring together, are prominent sources of ions which, in solution, contribute to the hardness of streams; however, sulfates, chlorides, and other compounds may also contribute to total hardness.

The combination of high carbonate concentration and low carbon dioxide content, characteristic of these streams, apparently constitutes a more favorable environment for plants and animals than do acid conditions-for, generally speaking, hard-water streams support a varied and abundant association of organisms. Most unpolluted major streams exhibit a pH value on the alkaline side of neutrality, and, indeed, tend toward uniformity of composition generally.

Although the relationships of carbon dioxide, bicarbonates, and carbonates will be considered a bit further with respect to dissolved solids, we might briefly summarize their activities in relation to pH in lakes and streams thus:

- The pH value varies inversely as the dissolved carbon dioxide concentration, and directly as the bicarbonate concentration;
- The critical value relating to the presence or absence of free carbon dioxide is pH 8, the free gas being absent above that value;
- The absence of free carbon dioxide does not limit photosynthesis of certain algae and higher plants, some being adapted for utilization of carbon dioxide from carbonates, usually resulting in very high pH values.

Methane, hydrogen sulfide, and other gases derived from breakdown of organic substance may be present in high concentrations under proper conditions in very slow streams and in stagnant regions of stream pools. In moving water, however, turbulence due to current tends to diffuse and eliminate the gases.

DISSOLVED GASES IN ESTUARIES

The considerable differences between the chemistry of fresh water and that of sea water bring about complex relations of dissolved substances in general within an estuary. The occurrence of dissolved materials relates to the disproportion of dissolved materials found in the waters of the inflowing stream at the upper end of the estuary and in the sea water at the mouth of the estuary. Between these two relatively uniform states there exists a considerable gradient in processes and conditions. Dissolved gases are distributed in an estuary in accordance with turbulence and current factors, biological activities, and salinity and temperature effects. We have previously considered most of these factors and their influence on dissolved gases with respect to lakes and streams; in general, the interrelationships are similar in estuaries. One new factor is quite influential in estuaries; this is salinity, the total amount of dissolved inorganic solids.

OXYGEN

In salt water the solubility of oxygen decreases as water temperature and salinity increase. If we visualize cool, fresh (low-salinity) water entering the uppermost reaches of an estuary and grading toward somewhat warmer and more saline sea water at the lower extreme, we can begin to appreciate the adjustments taking place throughout the length of the estuary. Table gives saturation (absorption) coefficient values of oxygen in water at various temperatures and salinities.

The values show not only that temperature is the most important factor in determining oxygen solubility, but also that salinity manifests considerable influence.

We see then that less oxygen can be dissolved in sea water than in fresh water. At saturation at 15°C, a liter of sea water (at 36 ‰ salinity) contains 5.8 cc oxygen per liter of water; a liter of fresh water at the same temperature holds 10.3 cc. The importance of these relationships lies not only in the aforementioned possibility of a considerable linear oxygen gradient in a mixing estuary. There is also the possibility of fluctuations associated with stream flood seasons and the inflow of large quantities of fresh water, or with dry seasons when tidal flow of sea water dominates.

In nonmixing estuaries salinity stratification during summer often results in conspicuous differences in dissolved oxygen content in deep and in surface waters. In parts of Chesapeake Bay during summer the oxygen concentration may range from 90 to 100 per cent saturation at the surface, while bottom waters show from 40 to 50 per cent saturation. The surfacebottom differential decreases upstream as the depth decreases. In these shallow zones there is probably more mixing than in the bay proper.

Table. Saturation Coefficient Values of Oxygen And Carbon Dioxide In Water

Salinity (‰)	Tempera ture (°C)					
	0°	12°	24°			
	O_2	CO_2	O_2	CO_2	O_2	CO_2
0 (fresh water)	49.24	1717	36.75	1118	29.38	782
28.9	40.1	1489	30.6	980	24.8	695
36.1 (sea water)	38.0	1438	29.1	947	23.6	677

The range of oxygen concentration in the bay reflects the activity in the lighted trophogenic zone and the relatively slow replacement of sea water in the lower level. It may well be that even with more rapid replacement the water from the sea might be poor in dissolved oxygen.

Along with vertical differences, oxygen characteristically varies diurnally and seasonally within estuaries. The ranges of such variations differ, depending upon the nature of the freshwater source, the morphology of the basin, and effects of tides. In deep, turbid estuaries lacking the con-tribution from an abundant bottom flora, diurnal oxygen pulses are apt to be relatively slight. In Chesapeake Bay the range for surface waters was found to be from about 85 per cent of saturation in early morning to about 115 per cent in late afternoon. Shallow, clear estuaries may contain bottom growths of algae on which oxygen bubbles may be seen. In waters such as these the minimum-maximum diurnal range may exceed 200 per cent. The effects of incoming sea water during a tidal cycle may serve to mask the local conditions. Seasonally, the oxygen dynamics in an estuary may be influenced by variations in river discharge, tides, and day length and biological effects. Surface waters in one part of Chesapeake Bay were found to have 143 per cent saturation with oxygen in April, and about 42 per cent in August. Bottom waters in the same area reached 133 per cent saturation in October, and 24 per cent in June.

THE CARBON DIOXIDE SYSTEM AND pH

The solubility of carbon dioxide in estuarine waters is

determined primarily by the amount of sea water mixing with fresh water, and secondarily by temperature. The importance of sea water as a solubility factor rests, of course, upon salinity. In Table we see that at any temperature the absorption coefficient value of carbon dioxide decreases as the salinity increases. The high solubility of carbon dioxide is due to its chemical reaction with water. Although some of the carbon dioxide in sea water is in the form of the free gas and as carbonic acid, much more is present as bicarbonate and carbonates.

This condition results from the fact that sea water contains alkaline radicals in excess of the equivalent acid radicals which, as we have seen, shift the carbon dioxide system toward the carbonate formation, thereby reducing the free carbon dioxide concentration. The presence of the excess bases—boric acid and its borates, carbonic acid and the carbonates—in sea water serves to buffer the water against great changes in pH that might develop from the addition of acids or bases. The pH of sea water at the surface is very stable, usually ranging between pH 8.1 and 8.3. Since carbon dioxide uptake is greater in the presence of excess base, and since river waters usually contain a lower concentration of excess bases than sea water, we should expect to find a lower content of free carbon dioxide in the mouth of the estuary than in the upper reaches. Similarly, because river waters are seldom buffered, the free carbon dioxide concentration and pH should be more variable in that part of the estuary dominated by stream influences. Streams transporting large quantities of humic material in colloidal suspension are frequently slightly acid.

Upon meeting sea water the colloidal particles are coagulated and the pH shifts toward the alkaline side of neutrality. East Bay, Texas, receives considerable runoff from organically rich salt marshes, and it was found that during summer the pH ranged from 6.9 throughout much of the bay to 7.8 near the mouth where it discharges into the Gulf of Mexico. Gulf waters during the same period gave a pH value of 8.0. In view of what we have learned of oxygen-carbon dioxide-pH relationships generally, we should expect the vertical, diurnal, and seasonal distributions of carbon dioxide and pH to operate acording to certain principles. Although few actual studies have been made of carbon dioxide in estuaries, the pattern of distribution is expected to be the reverse of that of oxygen.

By the same rules, pH values should vary inversely as the free carbon dioxide content and directly as the dissolved oxygen concentration. It in the relationship:

Salinity ‰ = 0.030 + 1.8050 × chlorinity ‰

Chlorinity is rather easily and accurately measured by silver nitrate titration, using potassium chromate as an indicator. Salinity can be determined by electrical conductivity, and from measurements of density obtained by the use of hydrometers. The salinity of open seas generally ranges between about 33 and 38‰, with the average being near 35‰. Since the average salinity. of soft fresh water is 0.0651‰ and of hard fresh water 0.30‰, it is apparent that high estuarine salinities are derived almost wholly from sea water, while the diluting effect of the influent serves to reduce the concentration of dissolved salts.

In general, the proportion of dissolved salts in estuaries resembles that of sea water, while the total concentration is variable along the axis of the estuary. In some instances, however, the concentration of inflowing fresh waters may be such as to modify the normal ionic relationships in estuaries. Where this occurs the result is usually an increase in the ratios of carbonate and sulfate to chloride, and of calcium to sodium over those of average sea water. The momentary salinity may be regarded as a function of the quantity and quality of inflowing and outflowing waters, rainfall, and evaporation. Since these factors may vary with season (in some instances rather drastically), the general structure of the estuary also shifts. Therefore, attempts to fit estuaries into schemes of classification are often difficult. We can, however, recognize that certain estuaries are typically, or on the average, more or less saline than others. For example, the average salinity of the estuary-Iike Laguna Madre of Texas nearly always exceeds that of sea water.

Index